GREEN THUMBS: Corn & Beans

SCIENCE WITH SIMPLE THINGS SERIES

10970 S Mulino Road
Canby OR 97013
Website: topscience.org
Fax: 1 (503) 266-5200

Oh, those pesky **COPYRIGHT RESTRICTIONS** *!*

Dear Educator,

TOPS is a nonprofit organization dedicated to educational ideals, not our bottom line. We have invested much time, energy, money, and love to bring you this excellent teaching resource.

And we have carefully designed this book to run on simple materials you already have or can easily purchase. If you consider the depth and quality of this curriculum amortized over years of teaching, it is dirt cheap, orders of magnitude less than prepackaged kits and textbooks.

Please honor our copyright restrictions. We are a very small company, and book sales are our life blood. When you buy this book and use it for your own teaching, you sustain our publishing effort. If you give or "loan" this book or copies of our lessons to other teachers, with no compensation to TOPS, you squeeze us financially, and may drive us out of business. Our well-being rests in your hands.

What if you are excited about the terrific ideas in this book, and want to share them with your colleagues? What if the teacher down the hall, or your homeschooling neighbor, is begging you for good science, quick! We have suggestions. Please see our *Purchase and Royalty Options* below.

We are grateful for the work you are doing to help shape tomorrow. We are honored that you are making TOPS a part of your teaching effort. Thank you for your good will and kind support.

Sincerely, Ron Marson

Purchase and Royalty Options:

Individual teachers, homeschoolers, libraries:

PURCHASE option: If your colleagues are asking to borrow your book, please ask them to read this copyright page, and to contact TOPS for our current catalog so they can purchase their own book. We also have an **online catalog** that you can access at www.topscience.org.

If you are reselling a **used book** to another classroom teacher or homeschooler, please be aware that this still affects us by eliminating a potential book sale. We do not push "newer and better" editions to encourage consumerism. So we ask seller or purchaser (or both!) to acknowledge the ongoing value of this book by sending a contribution to support our continued work. Let your conscience be your guide.

Honor System ROYALTIES: If you wish to make copies from a library, or pass on copies of just a few activities in this book, please calculate their value at 50 cents (25 cents for homeschoolers) per lesson per recipient. Send that amount, or ask the recipient to send that amount, to TOPS. We also gladly accept donations. We know life is busy, but please do follow through on your good intentions promptly. It will only take a few minutes, and you'll know you did the right thing!

Schools and Districts:

You may wish to use this curriculum in several classrooms, in one or more schools. Please observe the following:

PURCHASE option: Order this book in quantities equal to the number of target classrooms. If you order 5 books, for example, then you have unrestricted use of this curriculum in any 5 classrooms per year for the life of your institution. You may order at these quantity discounts:

2-9 copies: 90% of current catalog price + shipping.

10+ copies: 80% of current catalog price + shipping.

ROYALTY option: Purchase 1 book *plus* photocopy or printing rights in quantities equal to the number of designated classrooms. If you pay for 5 Class Licenses, for example, then you have purchased reproduction rights for any 5 classrooms per year for the life of your institution.

1-9 Class Licenses: 70% of current book price per classroom.

10+ Class Licenses: 60% of current book price per classroom.

Workshops and Training Programs:

We are grateful to all of you who spread the word about TOPS. Please limit duplication to only those lessons you will be using, and collect all copies afterward. No take-home copies, please. Copies of copies are prohibited. Ask us for a free shipment of as many current **TOPS Ideas** catalogs as you need to support your efforts. Every catalog contains numerous free sample teaching ideas.

ISBN 0-941008-49-5

CONTENTS

INTRODUCTION

TEACHING NOTES

REPRODUCIBLE MATERIALS

GETTING READY

Welcome to **Green Thumbs: Corn and Beans**. *Here is a checklist of things to think about, and preparations to make, before beginning your first lesson.*

✔ WHEN TO SCHEDULE

You may teach this unit during any season of the year. Corn and beans grow slower in winter because of cooler temperatures and reduced hours of daylight; faster in summer because of warmer temperatures and more daylight. Your plants will grow and develop more or less "by the book" if you maintain ambient temperatures in your classroom between 60° F and 80° F (15°C to 27° C). In winter, set the plants in direct sunlight whenever possible; in summer, avoid long exposure to direct sunlight.

Consult your school calendar before you begin. Start on Monday, followed by 5 weeks of 5 days each. If your schedule doesn't have that many continuous days without interruptions or vacations, it's easier to miss days near the end of this 5-week period, or during the first week, but never miss Friday of the first week. Make up each day missed by doubling up on the work, if possible. Otherwise our daily TOPS calendar of experiments will no longer correspond to your actual day in the week.

Don't attempt to teach this module on an alternating M/W/F or Tu/Th schedule. There is too much to do, to observe and to record. If you have an alternating science schedule, consider rearranging it: teach science on a daily basis for 5 weeks at the expense of some other subject; then make up that lost subject by teaching it at the expense of science over a similar time period.

✔ WHERE TO GROW

Full-spectrum diffused daylight that enters your room through windows is sufficient to grow corn and beans. Narrow-spectrum fluorescent lighting will not support normal photosynthesis. Exposure to direct sunlight, while beneficial, is not necessary. It may even be harmful in hot weather if the plants are not well-watered.

Any horizontal surface near a window in your classroom makes a suitable growing area. Table tops, ledges, shelves, even the floor will serve. All experiments in this unit fit on a single sheet of notebook paper. You'll need less than 1 square foot of growing space per lab group.

Estimate available growing space in your classroom by laying sheets of notebook paper, side by side, over all suitable growing surfaces. If you lay down more papers than the total number of students in your class, then you have sufficient space for everyone to do individual work. This will engage your students most effectively in the learning process, because each one gets to complete all phases of every experiment.

You can cut your space requirements in half by assigning students to work in cooperative pairs. All students are still responsible for keeping personal plant journals, but they pair up to work through lab instructions, and to share growing trays, balances, pole planters and numbered bottle caps.

Avoid, if possible, lab groups larger than two, or pairing dominant and passive students. Consider which students work most cooperatively together, then pair them by writing both names at the top of a set of lab pages (see next paragraph). If students prefer to work alone, let them do so if space permits.

✔ WHAT TO COPY

❑ **11 Lab Pages:** Find these reproducible student pages toward the back of this book. They are numbered lab page 1, lab page 2, ..., lab page 11, in the lower right corner of each page. Duplicate and collate these sheets in sequence, in 11-page sets, one for each lab group in your class, plus a few extras. Staple each set in the upper left corner (front and back to prevent back pages from working loose). Pencil in the name(s) of each student or lab group at the top of the first page. Don't use pen — you'll want to reuse these lab instructions next time you teach this unit. When you distribute these instructions on MON/-4, tell students they are for reference only. They should use them as they would any textbook, returning them unmarked and in good shape for others to use.

❑ **12 Journal Pages:** Find these following the lab pages at the back of this book. They are numbered journal page 1, journal page 2, ..., journal page 12, in the lower right corner of each page. Duplicate and collate these sheets in sequence, in 12-page sets, one for each student in your class, plus a few extras. Paper-clip (don't staple) each set in the upper left corner. Hand these out on MON/-4 as well. Students will ask for them soon after you have distributed the lab pages.

❑ **Take-Home Test:** Find this page just before the lab pages. Duplicate one for each student in your class, plus extras. Paper-clip these together and save until FRI/28, the last day of this unit.

✔ WHAT TO GATHER

Once you gather the simple materials listed on the next page, you're ready to begin an educational adventure — a wonderful integration of science, math, language and drawing.

Gathering Materials

Listed below is everything you'll need to teach this module. You already have many of these items. The rest are available from your supermarket or garden store. Keep this classification key in mind as you review what's needed:

special in-a-box materials:	general on-the-shelf materials:
Italic type suggests that these materials are unusual. Keep these specialty items in a separate box. After you finish teaching this unit, label the box for storage, and put it away to use again next year.	Normal type tells you these materials are common. Keep these basics on shelves readily accessible to your students. The next TOPS unit you teach will likely use many of the same materials.
(substituted materials):	*optional materials:
Parentheses enclosing any item suggest a ready substitute. These alternatives may work just as well as the original. Don't be afraid to improvise, to make do with what you have.	An asterisk sets these items apart. They are nice to have, but optional. They are probably not worth an extra trip to the store, unless you are gathering other materials as well.

All materials are listed in order of first use. Start gathering at the top and work down. Ask students to bring recycled items from home. Most of these materials will be used sooner rather than later, as students construct most lab equipment in the first few lessons.

Needed quantities depend on class size, and how you organize students into lab groups. Adjust your own numbers up or down as needed:

Q_1 / Q_2

- **Single Student:** Enough for 1 student to complete all the experiments. (Q_1)
- **Classroom:** Enough for 30 students when organized into 15 lab pairs. (Q_2)

KEY: *special in-a-box materials* / (substituted materials) / general on-the-shelf materials / *optional materials

Q_1/Q_2	Materials
1/30	pairs of good, pointed scissors
1/3	boxes paper clips, medium size, all the same brand
24/650	sheets notebook paper — college ruled preferred, wide ruled acceptable
1/4	rolls clear tape
1/1	stapler
1/3	spools thread
1/2	rolls masking tape
1/15	*metric rulers
2/30	straight plastic drinking straws
2/45	straight pins
1/15	wooden clothespins
1/15	empty cans, medium size, 14 -16 oz
1/1	roll aluminum foil
1/15	size-D batteries, dead or alive
1/1	pkg dry pinto beans, sold in grocery stores } a fresh supply each year ensures viability
1/1	pkg dry popcorn, sold in grocery stores } a fresh supply each year ensures viability
2/45	baby food jars, medium or large size
4/60	bottle caps from soft drink (or beer) bottles
1/15	plastic tubs, deli or margarine
1/4	rolls paper towels
1/15	cardboard milk cartons (styrofoam egg cartons) — see teaching notes for MON/10
1/1	water source — distribute watering bottles to avoid traffic jams
1/30	*hand lenses — see teaching notes for WED/5
.5/8	*quarts vermiculite, sold in garden and variety stores*
1/1	jar of petroleum jelly
1/1	bottle blue food coloring — use self-dispensing bottle or provide eye dropper
1/30	lids from baby food jars — used as food coloring palette, and for Take-Home Test
1/30	thick rubber bands
1/15	*large paper grocery bags — see teaching notes for MON/-4 and THU/27
1/1	**bag potting soil* — see teaching notes for THU/27
1/1	pkg dried lentils, sold in grocery stores — fresh for viability
1/1	pkg wheat berries — fresh, whole grains of wheat sold in health food or farm stores

Long-Range Objectives

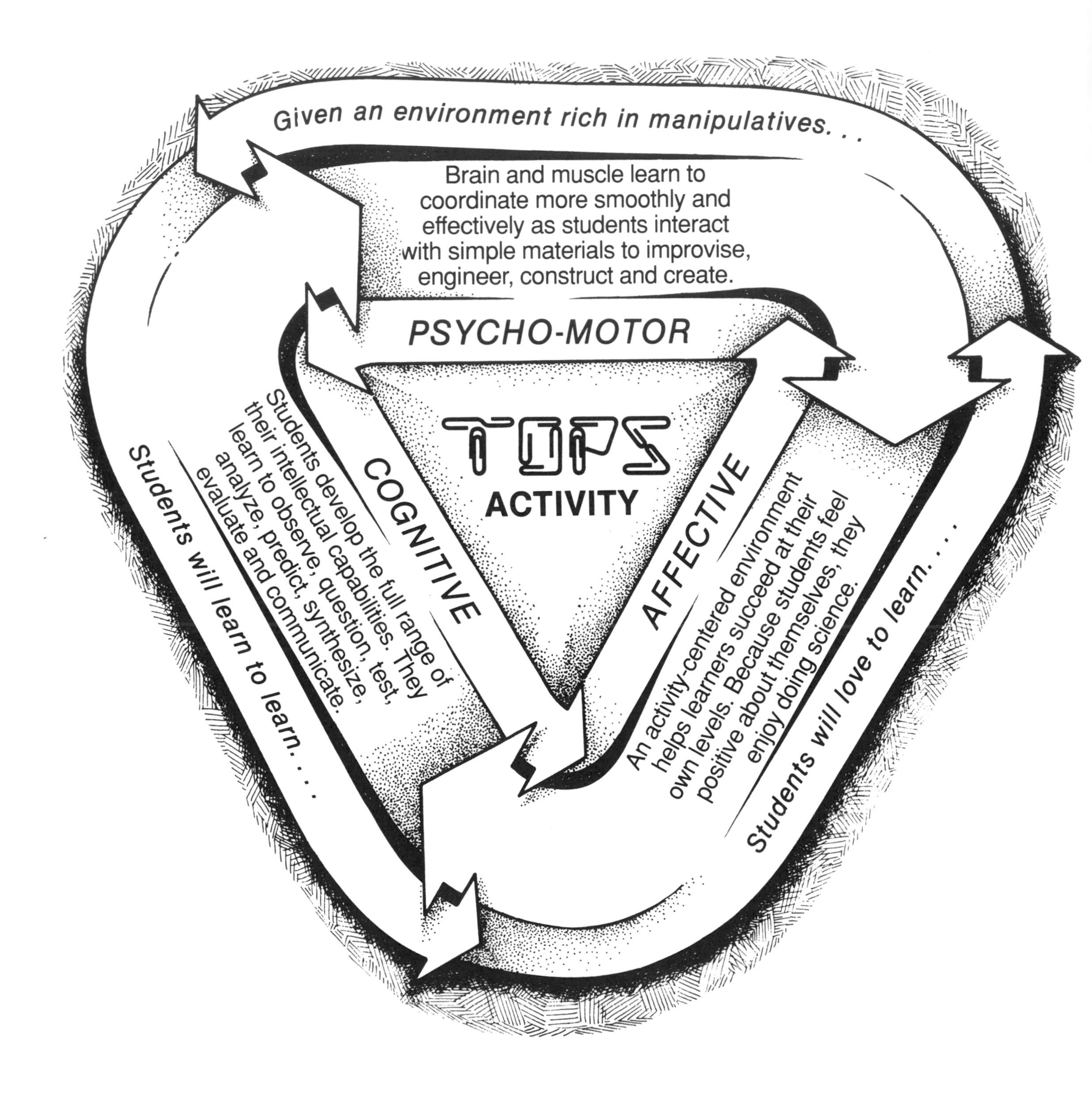

GAINING A WHOLE PERSPECTIVE

Science is an interconnected fabric of ideas woven into broad and harmonious patterns. Use the ideas presented below to help your students grasp the big ideas — to appreciate the fabric of science as a unified whole.

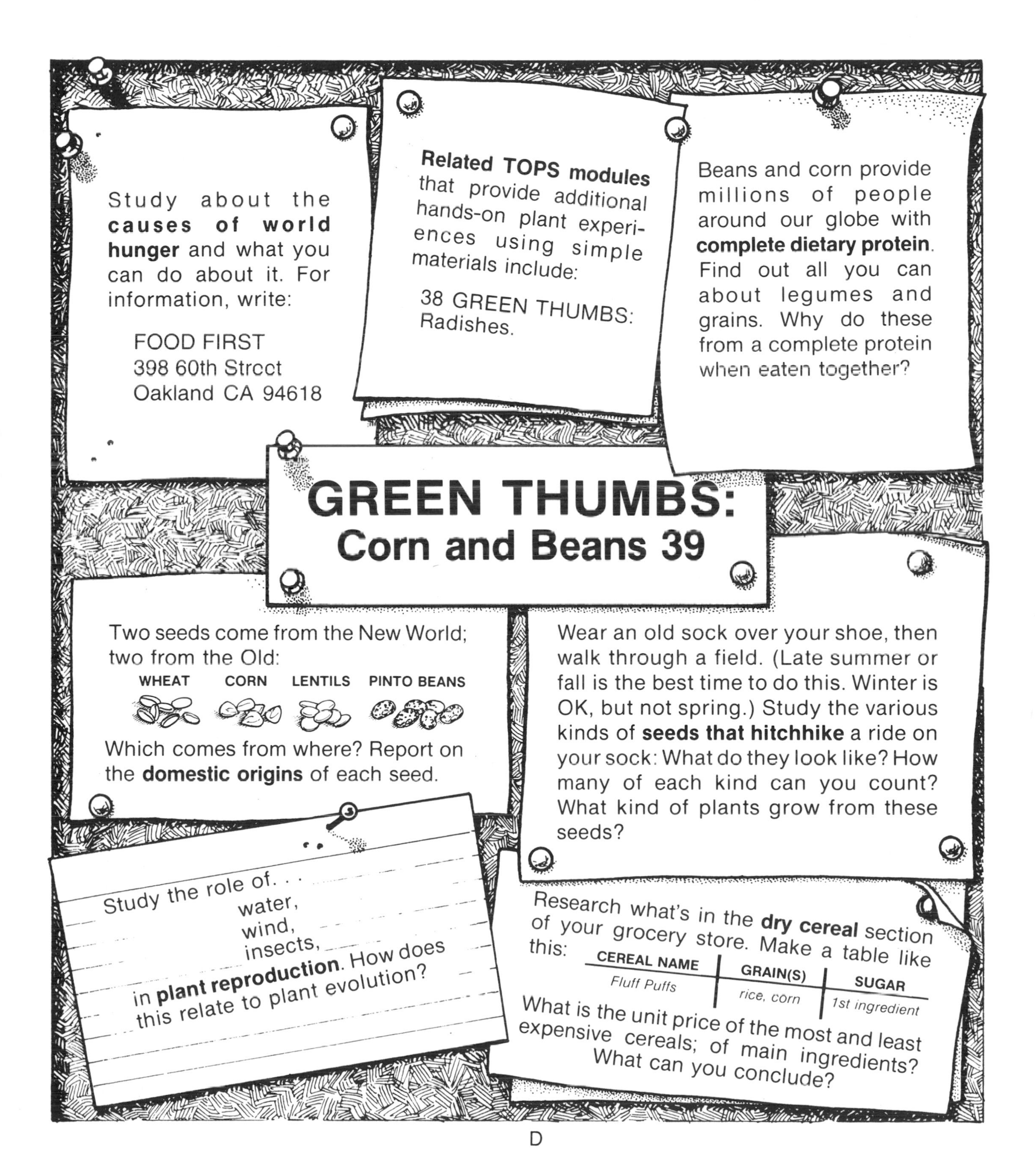

TEACHING NOTES

These teaching notes begin your first day's lesson on **Corn and Beans**. If you haven't already done so, begin with Getting Ready (page A) and Gathering Materials (page B).

Lessons in this book are organized by days of the week. The bold slash and number heading each lesson indicate the number of days that precede or follow "FRI/0," when the corn and bean seeds are initially exposed to water. This beginning lesson is numbered MON/-4 because it is scheduled 4 days before Friday's watering event. Next Monday's lesson is numbered MON/3 because a weekend plus a day will have elapsed since Friday's watering event. Lessons continue each weekday until FRI/28, four full weeks after FRI/0.

To complete this module on schedule, find a 5 week period in your school calendar that is relatively vacation-free. Because FRI/0 is "zero hour," it must not be missed by any lab group. The 4 lessons that precede FRI/0 can be reasonably condensed into 3 days, since no plants are growing during this introductory period. After FRI/0, school days are generally easier to miss and make up in later weeks of the schedule rather than in the earlier weeks.

The calendar below indicates those days that your students will be directed by lab instructions. These lab instructions alone will be used during the first calendar week. Through the last 4 weeks, your students will use a combination of both lab instructions and journal pages. Those days marked "(journal)" indicate days when *only* journal pages are used, without lab instructions.

Headings begin with a single or double asterisk. On this page a single asterisk precedes "Make your Plant Journal." This implies that individuals (every student) should make a plant journal. Turn the page to TUE/-3, and you'll see that a double asterisk appears in front of "Build a Balance." This means that pairs of students may work together to produce a single balance with shared materials.

2. This step is labor-intensive. To expedite the process, furnish every student with scissors. Borrow extras if necessary. Soon your room will be awash in strips of paper. Distribute extra wastebaskets, or brown paper bags, to collect the confetti.

After they finish cutting, remind your students to count all their separated pieces. If they count fewer than 19 weekdays and 4 paper masses, they have neglected to separate all pieces. (Journal page 9 is often missed because it separates into 3 parts.) If overlooked cuts are not discovered now, students will complain of missing pages during final journal assembly in step 5.

4. Each student will need 2 pieces of clear tape per cutout, or 38 total pieces. How do you cope with a classroom of students all needing so many pieces of tape at the same time? Easy: as students ask for tape, distribute strips as long as your arm (about 19 inches). Stick each long piece to the student's desk, with a paper clip under one end to make it easy to lift. Tell your students to snip short pieces from this longer strip, as wide as a fingernail (about ½ inch). They should temporarily stick these short pieces along the edges of their desks.

First Week:	**MON/-4**	TUE/-3	WED/-2	THU/-1	FRI/0
Second Week:	MON/3	(journal)	WED/5	THU/6	FRI/7
Third Week:	MON/10	(journal)	WED/12	(journal)	FRI/14
Fourth Week:	MON/17	TUE/18	(journal)	THU/20	FRI/21
Fifth Week:	(journal)	**TUE/25**	(journal)	(journal)	

* Make your Plant Journal — lab MON/-4

***START:* 1.** Get 12 journal cutouts from your teacher.

2. Cut around the dashed lines into 23 separate parts:

3. Paper clip the four paper masses together, and set them aside.

TEACHING NOTES

Tape each cutout flush with the top-right corner of the notebook paper, using 2 snippets of tape as shown. This keeps all extra paper to the right of the margin holes. In step 6, the stapler will only need to penetrate a manageable 20 sheets of paper to bind the journal.

5. Students should follow this sequence: MON/3, TUE/4, WED/5, ..., THU/27. Weekends, of course, are missing.

6. Before stapling, make sure all pages are pressed together evenly, forming a uniform spine. Back pages, in particular, have a tendency to fan off center and miss the staple. Because the margin of error between binder holes and paper edges is somewhat slim, direct younger students to bring their assembled journals to you for final stapling.

9. The paper masses tucked into this journal will be used tomorrow (TUE/-3). The journal itself won't be used again until next Monday (MON/3).

Photocopy Alternatives to Cutting and Taping

Alternative A: Load your photocopier with college-ruled notebook paper, then copy class sets of each of the 19 journal days for students to collate and staple. Some trial and error may be necessary to determine exactly where you should place cutouts on the copy glass so their images reproduces flush with the top right corner of the notebook paper. Blue lines on the copy paper will unavoidably show through each journal page reproduction, unless you try alternative B.

Alternative B: Draw the blue lines in darker on both sides of a piece of notebook paper, both front and back. Use a fine point pen so the lines distinctly reproduce without looking too dark. Make 20 reproductions per student of the *back* side of this lined master. Then run all of this paper through your copier again. This time reproduce the *front* side of your master with each cutout taped in place. If you do this accurately, margin "holes," front and back, should be superimposed.

Materials

☐ One set of photocopied lab instructions per lab group. See Getting Started, page A.
☐ Photocopied journal pages.
☐ Scissors, one per student.
☐ Grocery bags to collect paper trimmings (optional).
☐ Paper clips.
☐ Clear tape. Substitute glue, if you wish, as an alternative to taping.
☐ Notebook paper, 20 sheets per student.
☐ A stapler.

TEACHING NOTES

4-5. To make a balance that weighs true, it is critical that both arms have equal length. Help younger students, if necessary, to crease the precise center of their straw, and push a pin through this center in step 5.

5-12. These steps are not reproduced on this page for lack of space. (See lab pages 2-3 to reference these particular instructions.)

9. The straw will not balance perfectly level on its pin fulcrum until a tape rider is added in step 15.

10. You can dramatically reduce wasted foil by pre-tearing pieces off the larger roll. When students trace the dashed lines of the rectangle with the foil underneath, as directed, they will leave pencil marks on their reference-only lab instructions. This is OK.

15. Most balances should rest close to a level position. The tape rider, therefore, can be relatively small.

16. We have sized these paper masses for 20 pound bond or 50 pound book stock. If your copy paper is lighter or heavier than this, that's OK. As long as you reproduce all gram masses on a single grade of paper, they will remain internally consistent, accurate enough for all mass determinations in this book.

Caution students to cut out these paper masses with extra care. Any time the scissors stray off the line, they'll be subtracting mass from one standard while adding mass to its neighbor.

19. This step is less about getting the right answers (which can be recorded on scratch paper), and more about learning how to use the balance. If time permits, talk about mass variations and experimental error: Do seed masses vary more than paper clip masses? If you weigh the same seed twice, will you always get the exact same answer?

Typical result:

pinto bean = 370 mg
popcorn = 130 mg
paper clip = 750 mg

Your students will likely talk about "weighing" things in "grams." While such language is imprecise, it makes good sense while experimenting within the constant gravitational field of Earth. If your students aspire to becoming astronauts some day (or at least students of physics), you may wish to distinguish weight from mass: weight is the force that gravity exerts on an object; mass is a measure of inertia, or how much material an object contains, irrespective of the gravity acting upon it.

** Build a Balance

TEACHING NOTES

Materials

- ☐ Thread.
- ☐ Scissors.
- ☐ Masking tape.
- ☐ A plastic straw.
- ☐ A straight pin.
- ☐ Paper clips of equal mass and size.
- ☐ A wooden clothespin.
- ☐ A medium-sized can.
- ☐ Aluminum foil.
- ☐ A size-D battery, dead or alive.
- ☐ Paper masses. These were tucked inside each student's plant journal.
- ☐ A pinto bean seed and a popcorn seed.

TEACHING NOTES

So far, your students have made plant journals and balances. This lesson is devoted to constructing all remaining equipment needed to study corn and beans: a pole planter, a numbered deli-tray, and a storage mat that organizes all equipment in one place.

1-3. Wrapping the paper clip in tape, and sandwiching it between layers of sticky masking tape, provides a firm anchor for the long bean pole constructed in steps 4-13.

5. Use notebook paper that has the same line width as the plant journal. Mixing a "college-ruled" pole with a "wide-ruled" journal, for example, while no big deal, enlarges the plant drawings a little beyond actual size.

6. It is essential to fold the notebook paper *in half* 4 times in succession, forming 16 layers. This yields a stiff, narrow "pole" about 1 finger wide.

10. Students habitually number papers *above* the lines. It will take extra explaining on your part to get them to depart from this norm and number *through* the lines. Remind students to number *up* from the bottom, and to count the seam between the paper tubes as 1 line.

** Make a Pole Planter

lab WED/-2

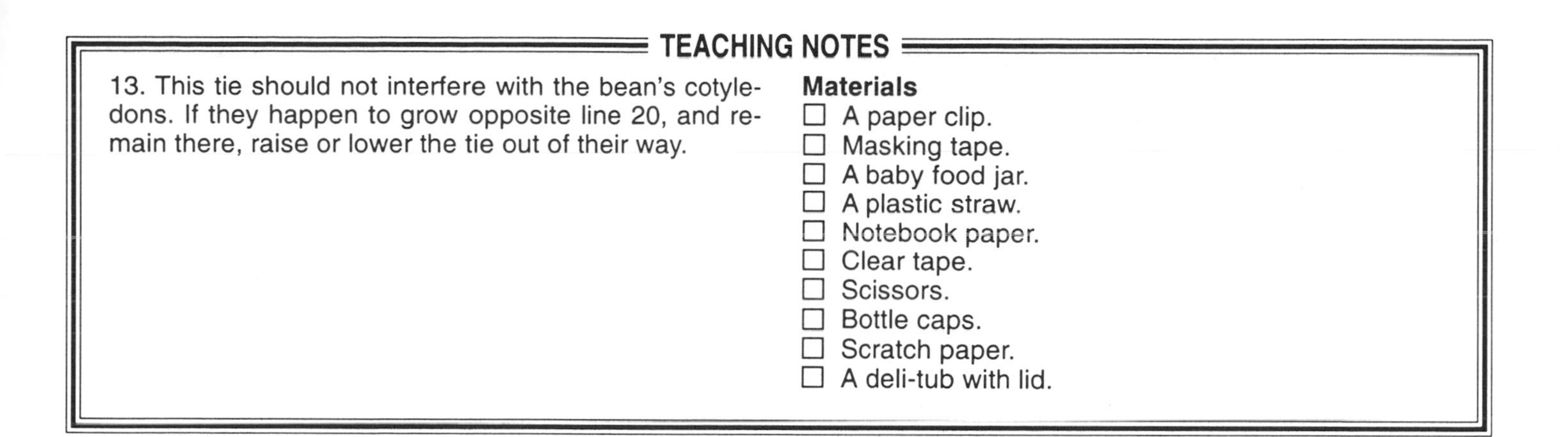

TEACHING NOTES

13. This tie should not interfere with the bean's cotyledons. If they happen to grow opposite line 20, and remain there, raise or lower the tie out of their way.

Materials

- ☐ A paper clip.
- ☐ Masking tape.
- ☐ A baby food jar.
- ☐ A plastic straw.
- ☐ Notebook paper.
- ☐ Clear tape.
- ☐ Scissors.
- ☐ Bottle caps.
- ☐ Scratch paper.
- ☐ A deli-tub with lid.

TEACHING NOTES

Today your students will draw a bean map. It is the only project they complete this week that does not involve building lab equipment. This map is an important skill-building exercise. It involves mental rather than physical preparation, for the many seed, sprout and plant drawing they will complete in the weeks that follow.

If you demand high quality output now, in this initial drawing exercise, you will set a standard of excellence that students can aim for in all remaining journal drawings. Observational skills, concentration, patience and attention to detail will all improve. Expect the best from your students, and they will deliver their best. Haphazard two-minute scribbles are not acceptable. Ask for beans that look as photographic as possible within each student's skill level.

1. Carefully trace both the square and the 2 bean outlines inside.

3. A small percentage of pinto beans have seed coats that are unusually dark. A few may be especially small or wrinkled. The idea here is to *avoid* unusual looking beans so students will have to rely on the accuracy of their bean maps (step 4) to find each special bean in a larger group (step 5-6).

* Make a Bean Map

lab THU/-1

TEACHING NOTES

4. Allow at least half an hour to complete this step. Good representational drawings require a slow, deliberate hand that draws exactly what the eye sees. Praise those who carefully capture the intricate seed patterns on their special bean.

Materials

- ☐ Scissors.
- ☐ Pinto beans.
- ☐ Clear tape.

TEACHING NOTES

2. Hand strength, good eye-hand coordination and a sharp pair of heavy duty scissors are all needed to make neat cuts along 3 edges of the milk carton. Assist younger children as necessary.

3-6. The crossing of these diagonal lines marks the center of the seed tray's bottom. When vermiculite and water are added later, this point will be close to the tray's true center of gravity.

4-6. Both paper towels should fit snugly through this thumb hole, not too tightly nor too loosely. An excessively tight fit may constrict the capillary flow of water from jar to tray, retarding overall plant growth in the weeks that follow. An excessively large hole will permit water to drain off too quickly when corn and bean sprouts are pushed into a soft slurry of vermiculite on Mon/10. This can be corrected by firming up the hole with extra paper towel. Push it into the soft center depression like a plug.

10. Remind students to keep the lids on their seed trays closed when not observing their seeds. This reduces cooling by evaporation and speeds germination.

11. Seeds are planted. Improvised lab equipment is all arranged and at the ready. What will the corn and bean seeds look like after a weekend of germination? Anticipation should be building and curiosity running high. Expect near perfect attendance on MON/ 3!

Materials

☐ A cardboard milk carton. Half gallon sizes are best, though smaller sizes down to 1 quart will also work. Rinse and allow to drip dry before using.
☐ Scissors. These should be heavy duty, capable of cutting through the coated cardboard of milk cartons. Blunt-nosed children's scissors are not up to the job.
☐ Masking tape.
☐ Soft kitchen-grade paper towels.
☐ Wide mouth jars. Pint or quart sizes are suitable.
☐ Pinto bean seeds.
☐ Popcorn seeds.

** Make a Milk Carton Seed Tray — lab FRI/0

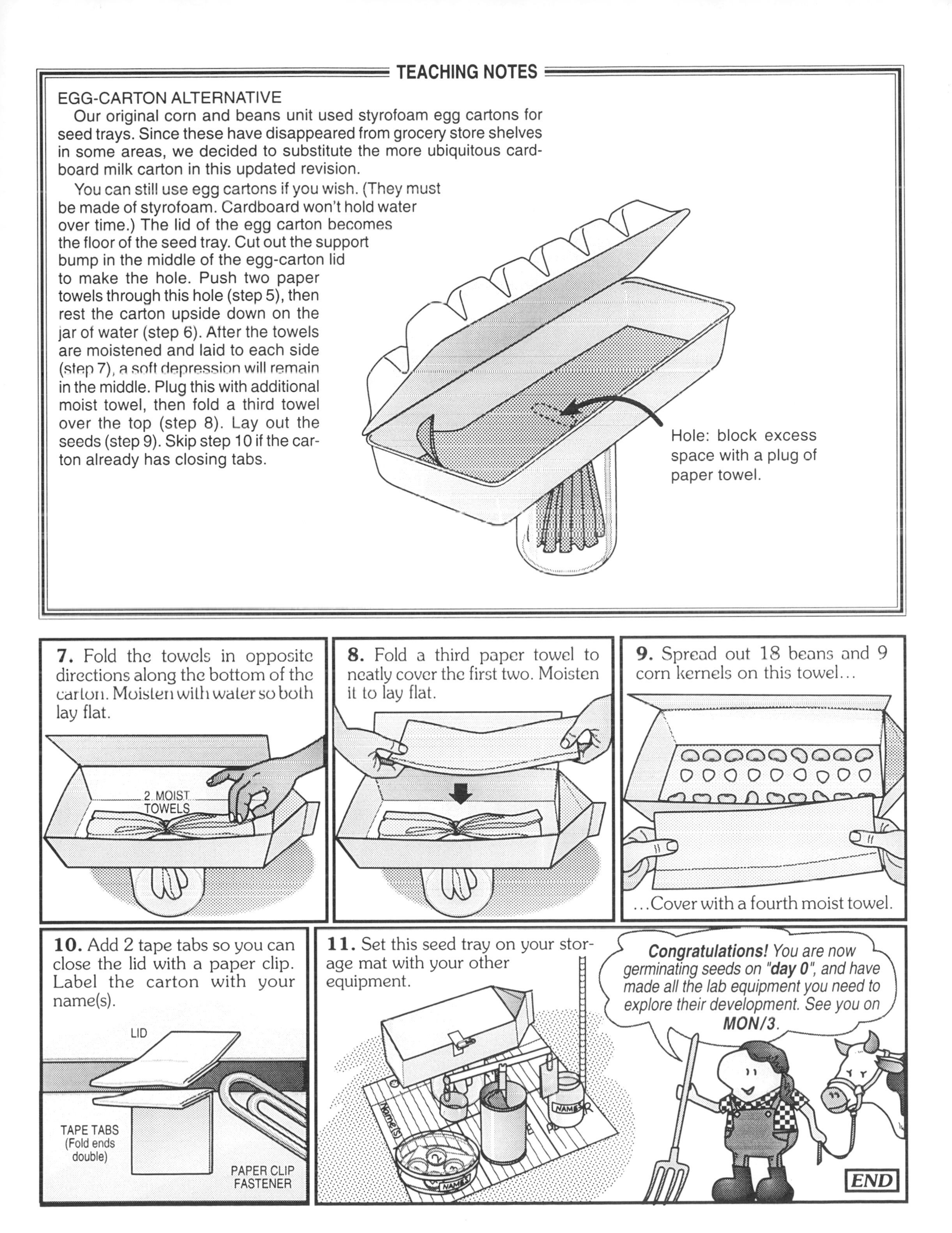
TEACHING NOTES
EGG-CARTON ALTERNATIVE
Our original corn and beans unit used styrofoam egg cartons for seed trays. Since these have disappeared from grocery store shelves in some areas, we decided to substitute the more ubiquitous cardboard milk carton in this updated revision.
You can still use egg cartons if you wish. (They must be made of styrofoam. Cardboard won't hold water over time.) The lid of the egg carton becomes the floor of the seed tray. Cut out the support bump in the middle of the egg-carton lid to make the hole. Push two paper towels through this hole (step 5), then rest the carton upside down on the jar of water (step 6). After the towels are moistened and laid to each side (step 7), a soft depression will remain in the middle. Plug this with additional moist towel, then fold a third towel over the top (step 8). Lay out the seeds (step 9). Skip step 10 if the carton already has closing tabs.
Hole: block excess space with a plug of paper towel.
7. Fold the towels in opposite directions along the bottom of the carton. Moisten with water so both lay flat.
2 MOIST TOWELS
8. Fold a third paper towel to neatly cover the first two. Moisten it to lay flat.
9. Spread out 18 beans and 9 corn kernels on this towel...
...Cover with a fourth moist towel.
10. Add 2 tape tabs so you can close the lid with a paper clip. Label the carton with your name(s).
LID
TAPE TABS (Fold ends double)
PAPER CLIP FASTENER
11. Set this seed tray on your storage mat with your other equipment.
NAME(S)
NAMES
Congratulations! You are now germinating seeds on "day 0", and have made all the lab equipment you need to explore their development. See you on MON/3.
END

TEACHING NOTES

These teaching notes (and all that follow) correspond visually to the pages in each student's plant journal. Model answers always appear in bold type.

Before your class begins today's activity, review the function of the numbered box at the top of each journal page. Point out that it acts as a road map when navigating between lab directions and written journal responses. Today, on MON/3 for example, you begin with four lab steps followed by a transition, in step 4, to three journal questions. Tomorrow, TUE/4, will be devoted exclusively to journal responses with no lab instructions at all.

2. These seeds will soften as they absorb water over the next 2 days, making them suitable for dissection on WED/5.

3. This step marks the beginning of a 4-week tradition. Every day your class will sketch their most advanced bean and corn plant. These are now in the form of sprouts, and they will rapidly grow into plants. Hold your students to their "bean-map" standard of excellence established on THU/-1. They'll reward you (and themselves) with their best effort. Soon they'll have an entire journal filled with excellent work, something to take home and proudly show family members.

Typical development, shown actual size:

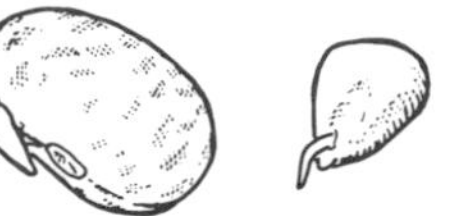

5-6. Vertical arrows, at the bottom of this journal page, and on many pages that follow, show where long, straight lines should be drawn downward to function as space dividers. Students should do this now if they missed that instruction when they first assembled their journals.

Ask your students to bring their completed journal pages to your desk for your initialed approval. For details, turn one page to teaching notes TUE/4.

Additional Materials

Your class has already constructed the lab equipment they'll need to complete most remaining activities. The word "additional" on our materials heading above indicates that from here forward, we will reference only new items, not already assembled on each lab group's storage mat. If no materials are listed, nothing new is required.

☐ Pinto beans and popcorn.

* Begin your Journal — lab MON/3

lab	1	2	3	4		
journal				4	5	6

journal MON/3

4. The **BEAN** is covered by a speckled **seed coat**. It protects the plant embryo, wrapped inside, from insects and water loss. The scar where the bean was once attached to its pod is called the **hilum**. Next to the hilum is a tiny hole called the **micropyle**. The embryo absorbs water most rapidly through this opening.

The largest part of the **CORN** is a yellow fruit called the **endosperm**. It is a food source for the white **embryo**, living inside, that will grow into a corn plant.

FIND each **bold** word on a real seed.
LABEL it on the line.
DEFINE what it does.

Seed coat:
A covering for the bean embryo: it keeps water in and insects out.

micropyle:
A small hole that allows water to pass rapidly into the bean embryo.

hilum:
The scarred point of attachment of the bean to its pod.

embryo:
The living part of the corn seed that grows into a plant.

endosperm:
A food storehouse for the corn embryo.

DRAW LINES DOWN AT ALL ARROWS

5. List below 4 ways your dry bean and corn seeds are **different**.

6. List below 4 ways your dry bean and corn seeds are **similar**.

END

1. The bean is more rounded, the corn more angular.

2. The bean is colored in shades of brown, the corn in shades of yellow.

3. The bean is larger than the corn.

4. The bean embryo is hidden from view, but the corn embryo is partly visible.

1. Both seeds are covered by a hard, smooth coat.

2. Both seeds have white points of attachment.

3. Both seeds contain living embryos surrounded by a source of food.

4. Both seeds have the ability to germinate and grow into plants.

TEACHING NOTES

Consider requiring a daily check point to enforce high performance standards. If you do this now, you'll avoid long hours of journal grading later. Your initials at the top of each day's journal page could signify that each page has passed inspection. At the end of this unit, when it comes time to grade the entire journal, a quick count of all the pages you've initialed will enable you to easily quantify a final grade.

Tell your students how you plan to grade them. If they perceive that your initials are valuable, and you tell them that you only initial current work, they'll seek you out for daily work reviews, not the other way around.

3. Mass of beans; mass of corn: Mass is a measure of inertia, the amount of matter in an object. Weight is a measure of how strongly that object is attracted by gravity. Nontechnical English does not adequately distinguish between these terms. To be precise, a 750 g seed is "massier," but not necessarily "heavier," than a 370 g seed. In outer space, both seeds would weigh nothing, even though their masses are unchanged.

4-DAY-OLD SPROUTS:

BEAN

CORN

lab journal 1 2 3 4 5

journal TUE/4

* **Draw your Sprouts**

***START:* 1.** Draw your fastest growing bean and corn sprout on the left page.

2. Unwrap your 10 seeds soaking in the wet towel since MON/3. Set dry corn and bean seeds beside them on your table top.

DRY SOAKED

3. Describe these before/after changes, using *complete* sentences:

DRY SEEDS ...**before** adding water	**SOAKED** SEEDS ...**after** 24 hours
color of beans: ***The seed coat is tan, covered with dark brown, irregular spots.***	color of beans: ***Light and dark brown patterns remain on the seed coat, but they have faded.***
color of corn: ***A deep yellow, translucent endosperm surrounds a nearly white embryo.***	color of corn: ***The endosperm is now a lighter yellow and more opaque.***
texture of beans: ***It feels hard and smooth.***	texture of beans: ***It feels soft and slightly sticky; when squeezed, water appears.***
texture of corn: ***It feels hard, with well-defined angles and ridges.***	texture of corn: ***It still feels hard, but the surface is rounder and smoother.***
mass of beans: ***It is heavier than the corn, with a mass of 370 g.***	mass of beans: ***Water makes it much heavier than before, with a mass of 750 g.***
mass of corn: ***It is lighter than the bean, with a mass of 130 g.***	mass of corn: ***Water makes it somewhat heavier than before, with a mass of 170 g.***
size of beans: ***It is bigger than the corn, with a length of 13 mm.***	size of beans: ***Water has swollen the bean to a length of 16 mm.***
size of corn: ***It is smaller than the bean, with a length of 9 mm.***	size of corn: ***Water has swollen the corn to a length of 10 mm.***

mm: 0 10 20 25

4. A good picture is worth a thousand words. Draw the changes you have described above in each box.

DRY BEAN | SOAKED BEAN | DRY CORN | SOAKED CORN

5. Return your 10 moist seeds to the seed tray.

END

TEACHING NOTES

2-4. All 5 beans and 5 corn wrapped in the damp towel are available for dissection. Students may inadvertently damage some of these in the process of learning how to open them cleanly. Corn is especially difficult to split open, because it has only 1 cotyledon.

5. In step 4 of the journal page MON/3, the white part of the corn seed was called the "embryo." Here it is called "one cotyledon." Actually, the single cotyledon surrounds the embryo so that both structures occupy the same white area. The corn embryo is difficult to see here because it has been torn apart. Both structures are easiest to distinguish in a soaked, unopened corn seed.

7. Unused seeds and fragments can be discarded or returned to the covered moist seed tray, as students wish. If returned, students may be surprised to notice that even half seeds sometimes grow — a poignant demonstration of life force.

Additional Materials

- ☐ A straight pin.
- ☐ A magnifying lens (optional).

* Look Inside

lab WED/5

2. Unwrap the bundle of seeds in your tray. Select an <u>un</u>sprouted bean and corn.

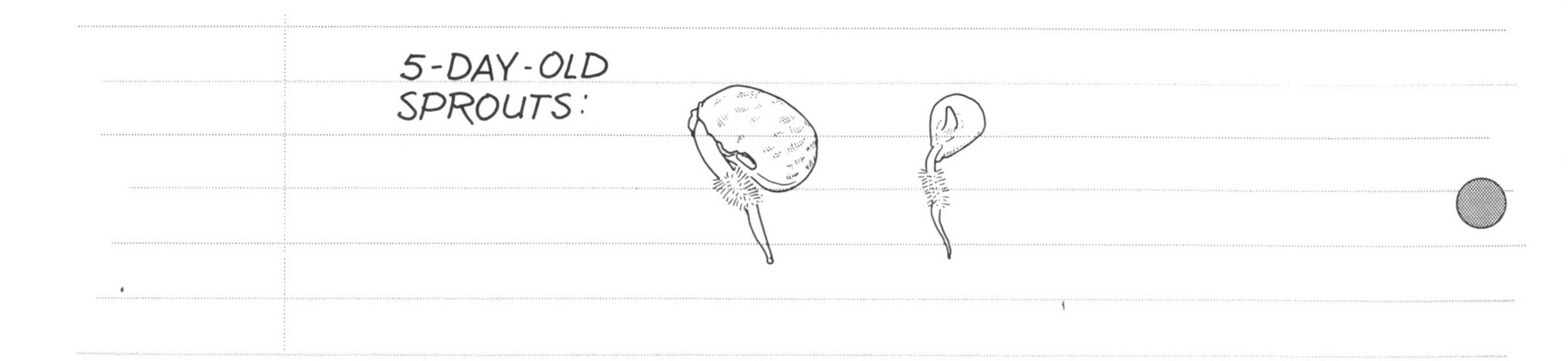

lab		2	3	4	5		
journal	1				5	6	7

* **Draw your Sprouts**

START: **1.** Draw your fastest-growing bean and corn sprout on the left page. *LAB*

5. The **BEAN** easily breaks apart into **two cotyledons**. These store starch and protein for the developing embryo, enabling its **plumule** to grow into the first 2 true leaves and its **radicle** to develop into the roots and lower stem.

The **CORN** does not divide easily because it has only **one cotyledon**. This structure, surrounding the embryo, absorbs starch and protein from the yellow **endosperm**.

FIND each **bold** word on a real seed.
LABEL it on the line.
DEFINE what it does.

radicle :
Becomes the roots and lower stem; a part of the embryo.

plumule :
Becomes the first 2 true leaves: a part of the embryo.

one cotyledon :
Absorbs food from the yellow endosperm for growth and development of the embryo.

two cotyledons :
Store starch and protein for growth and development of the embryo.

endosperm :
A food reserve of starch and protein; absorbed by the cotyledon to feed the embryo.

"MONO" means "ONE"...
"DI" means "TWO."

6. Which seed above is called a "monocot" (monocotyledon)? Explain below.

The corn is a monocot, because it has only 1 cotyledon.

7. Which seed is called a "dicot" (dicotyledon)? Explain below. *END*

The bean is a dicot, because it has 2 cotyledons.

TEACHING NOTES

2-4. This rolled up paper towel will hang just over the lip of a large baby food jar (as shown), farther down the side of a medium sized jar. It functions as a water sensor. When it wicks up enough water to feel cool and damp to the touch, the vermiculite inside holds sufficient water, as well. In all but the driest climates, there is a margin of safety of several days between the time the paper towel first dries out and the vermiculite inside the jar finally dries out. Your students can sense this with fingertips, or see the moisture on their skin. Even better, suggest that they use the lower lip, which has an especially high density of sensory receptors, and as a consequence is especially sensitive to temperature and moisture.

Vermiculite is easily compressed. The jar should be filled to its brim, but never packed down. Over time, some settling will occur naturally.

4-5. Hold the jar nearly upside down so all excess water drains away. Vermiculite won't fall out of the jar because it is held together by the strong cohesion of water. If the pencil holes in step 5 backfill with water, drain the jar more thoroughly.

6. The top of the bean and corn should just barely poke above the surface. Even though these plants may not grow significantly over the next 24 hours, students can still draw the exposed tops of the seeds when they sketch their pole planters for the first time on FRI/7. Roots that are too long to fit the hole can be trimmed back, without significantly slowing plant growth.

Additional Materials

- ☐ A thin rubber band.
- ☐ Vermiculite.

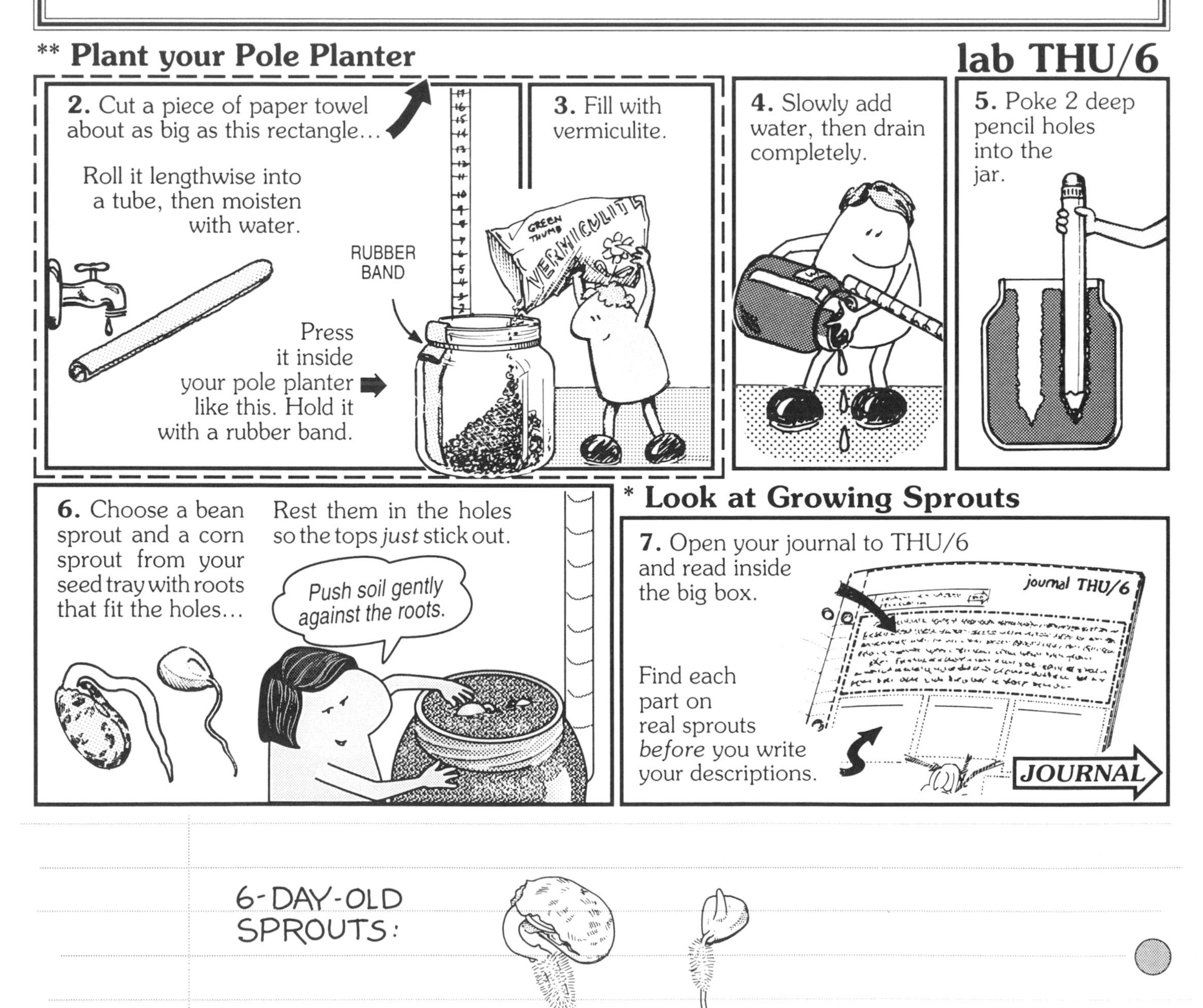

lab		2	3	4	5	6	7
journal	1						7

journal THU/6

* **Draw your Sprouts**

***START:* 1.** Draw your fastest-growing bean and corn sprout on the left page. *LAB*

7. The lower half of the **BEAN** radicle develops into a complete root system: The **primary root** first grows down into the ground seeking moisture. The hard **root cap** at the tip protects it from being torn apart as it pushes through the soil. Tiny tubes called **root hairs**, visible along the sides, absorb extra water. These are soon replaced by branching **secondary roots** that absorb water and anchor the plant firmly in the soil. With the root system in place, the **hypocotyl** directly above it lengthens into a lower stem, pulling its **two cotyledons** up through the soil into open sunlight.

Corn also develops a **primary root**, **root cap** and **root hairs** similar to the bean. Then the **coleoptile** pushes up from its white **cotyledon** surrounded by the yellow **endosperm**. This coleoptile forms a hollow protective tube that encloses the leaves until they grow above ground. Meanwhile, **adventitious roots** grow from the base of the coleoptile to further support the plant and absorb water.

FIND it on a real sprout; LABEL it on the line; DEFINE what it does.

hypocotyl*:** ***Forms the lower stem; pulls the cotyledons above ground as it lengthens.

secondary roots*:** ***These branch out from the primary root, replacing the root hairs. They absorb moisture and anchor the plant.

root hairs*:** ***Numerous tiny tubes on the primary root that absorb extra water.

two cotyledons

primary root*:** ***The first root to grow down; absorbs water.

root cap*:** ***A hardened shield that protects the tip of the primary root as it pushes down.

adventitious roots*:** ***These branch out from the base of the coleoptile. They absorb moisture and anchor the plant.

coleoptile*:** ***Wraps the leaves like a protective sheath until they grow above ground.

root hairs

primary root

cotyledon

endosperm

root tip

END

* Prepare your Journal for Daily Drawings

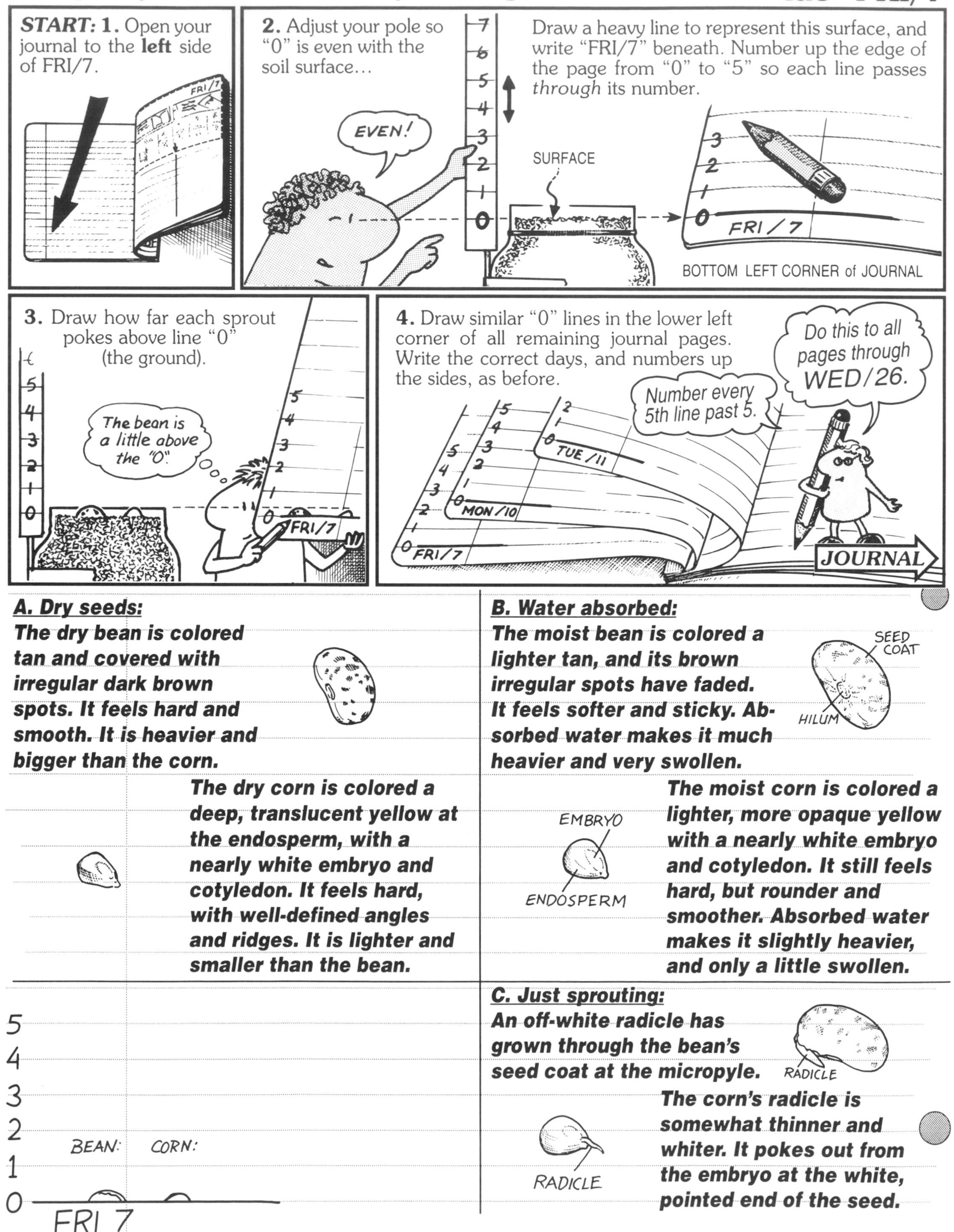

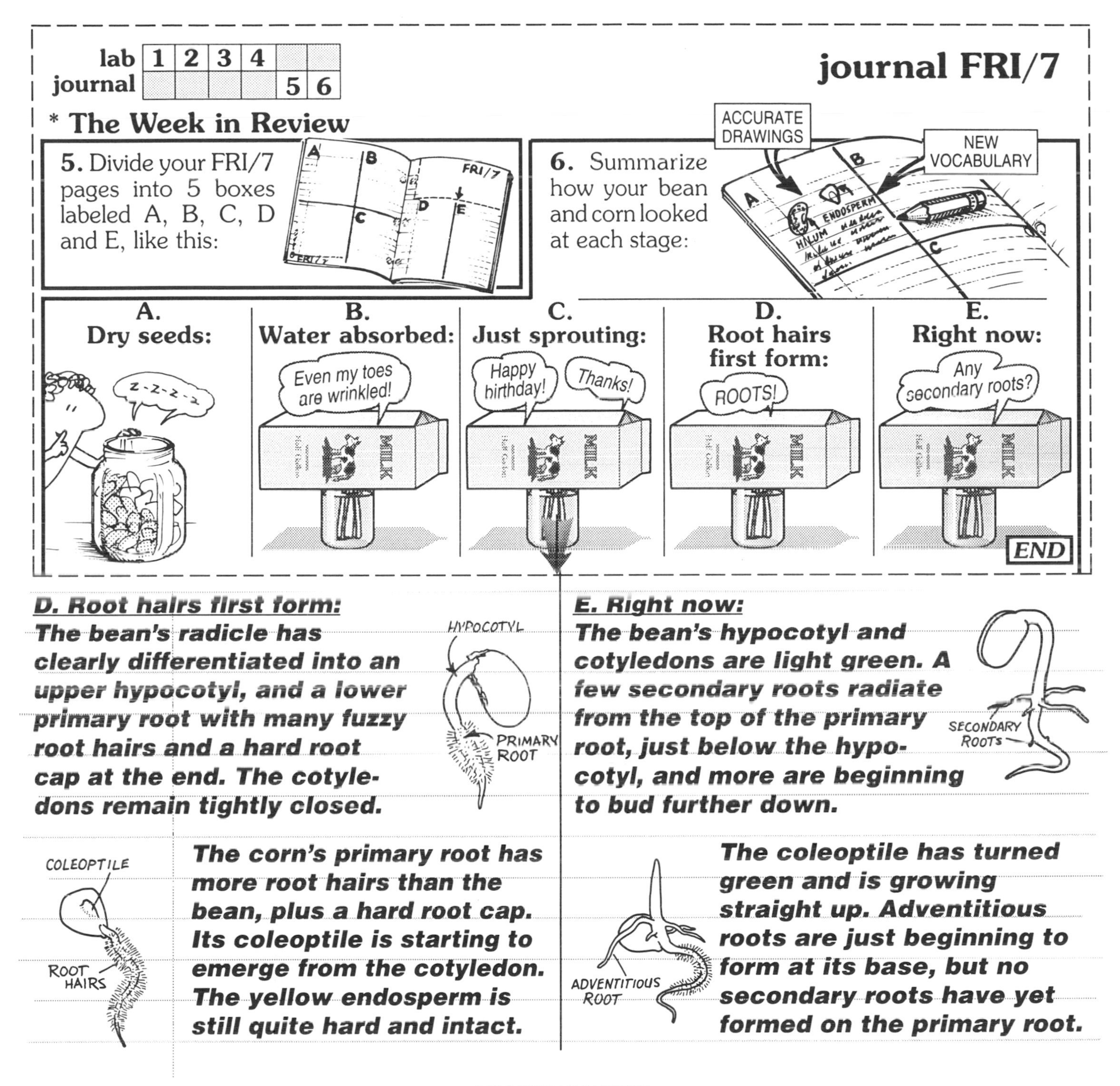

TEACHING NOTES

3. If you expect your classroom to cool below 60° F over the weekend, direct each lab group to construct a temporary greenhouse: invert a second baby food jar over the first, and stick masking tape around the perimeter where the two jars meet. This will aid growth as it reduces cooling by evaporation.

4. Prepare bottom left corners of the next 13 left-hand journal pages in this manner. Remind students to take their time and number each page neatly.

The bean will rapidly grow beyond 5 lines tall. Students should number higher lines (every 5th line will do), to keep pace with the growing bean.

5-6. These steps are an exercise in reexamining past notes to synthesize main ideas. They implicitly teach that journal notes are important; that they must be recorded accurately to make sense later. Require the best output your students can deliver — complete sentences and neat, labeled diagrams. You may need to demonstrate the use of simple straight lines to link words and pictures in labeled sketches.

** Plant your Seed Tray

2. The bean's hypocotyl is hooked so it can pull its 2 cotyledons "backward" out of the soil without breaking them open.

The corn's coleoptile pokes up through the soil like a spear, protecting the fragile leaves inside until they can grow into air and sunshine above ground.

TEACHING NOTES

2. If the bean's hypocotyl and the corn's coleoptile have not yet emerged, skip this question for now. Return to it in a day or two when the plants have developed as shown.

4. The tape tabs used to fasten the lid closed can now be removed.

5. This general housecleaning prevents mold. If you spot any traces of it in either the tray or jar, disinfect with a bleach solution and rinse thoroughly.

6-8. Add just enough water to float the vermiculite. This creates a liquid slurry that is thin enough to receive the tender corn and bean roots, yet thick enough to keep the plants standing upright once planted. While planting, adjust the thickness of this slurry as needed: thin it by adding more water; thicken it by draining water away.

Some roots may break as students pull each plant free of the towel, and the new transplants may stand somewhat askew in their new "soil." Not to worry. They will rapidly develop a new supporting root base and reorient upward growth toward the brightest light source in the room.

7. The new seeds introduced here, and again in step 2 of WED/12, insure a continuous supply of new sprouts in various stages of development for later experiments.

Consider planting healthy leftover sprouts in a reserve plant tray of moist vermiculite. Use these as insurance against future crop loss. Discard all infertile seeds and runts to prevent mold.

EGG CARTON ALTERNATIVE

If you are using styrofoam egg cartons as an alternative to milk cartons, direct students to remove the lids (the egg cup section), then clean the trays and transplant sprouts as directed. To support the growing plants, tape 4 plastic straws to the sides. Link these overhead with 2 more straws joined by paper clips and tape.

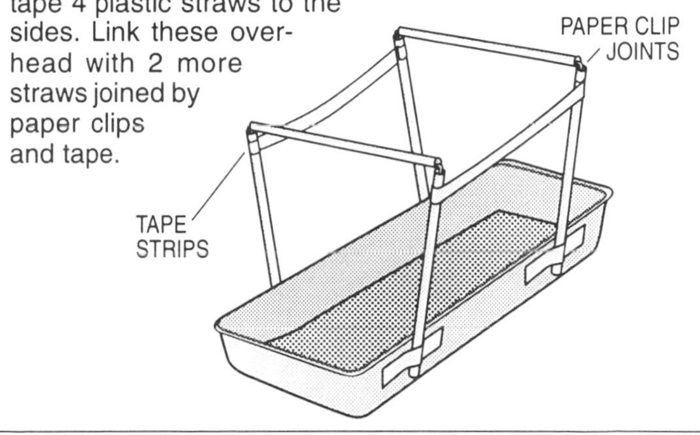

TEACHING NOTES

2. Before you leave your classroom tonight, write this notice on your blackboard as a reminder for tomorrow's work.

> STOP!
> Don't move anything.
> First answer question 2D
> from yesterday,
> before beginning today's work.

Ask students to observe other corn and bean plants, besides their own. Have they all grown into new orientations with respect to the light?

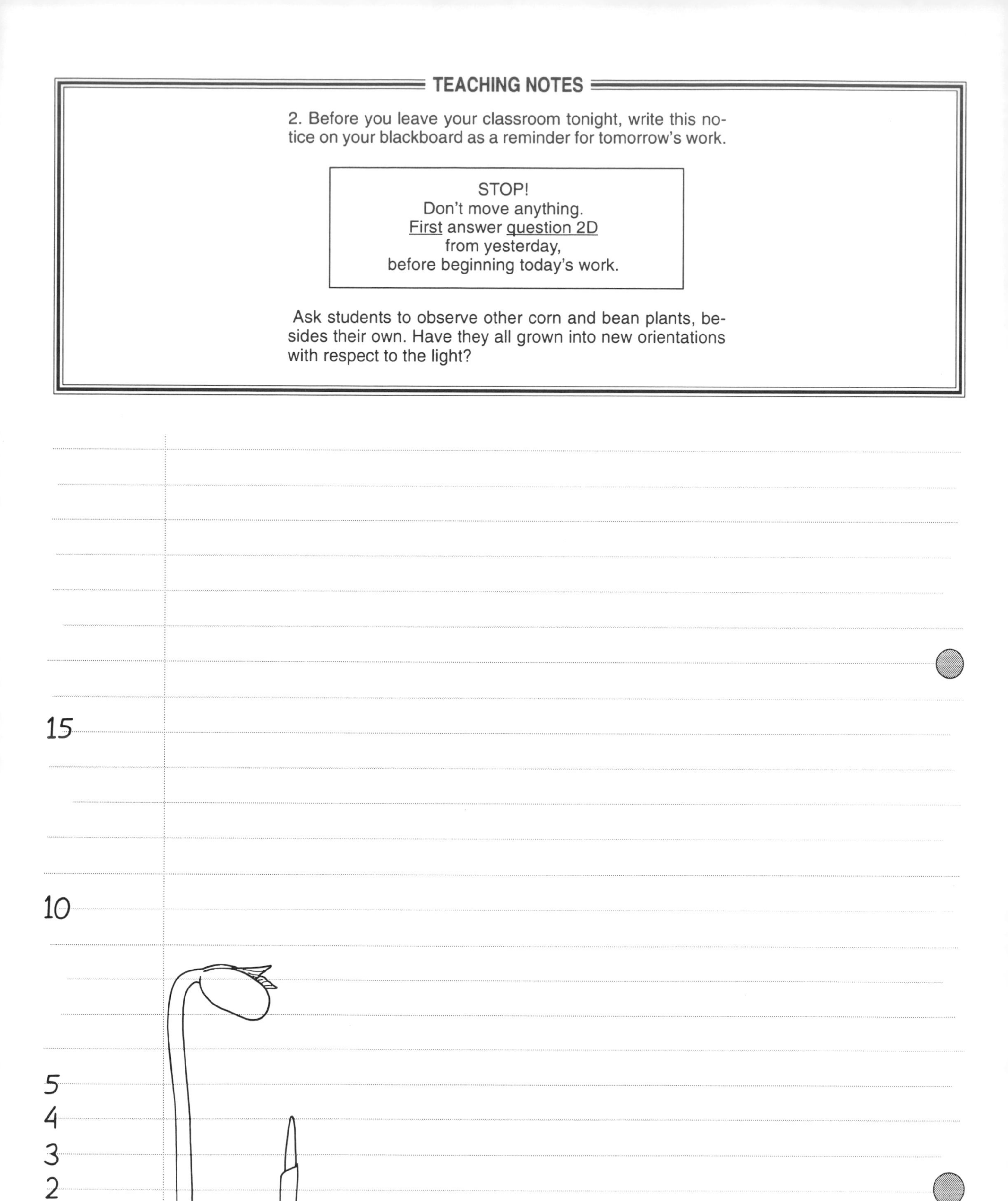

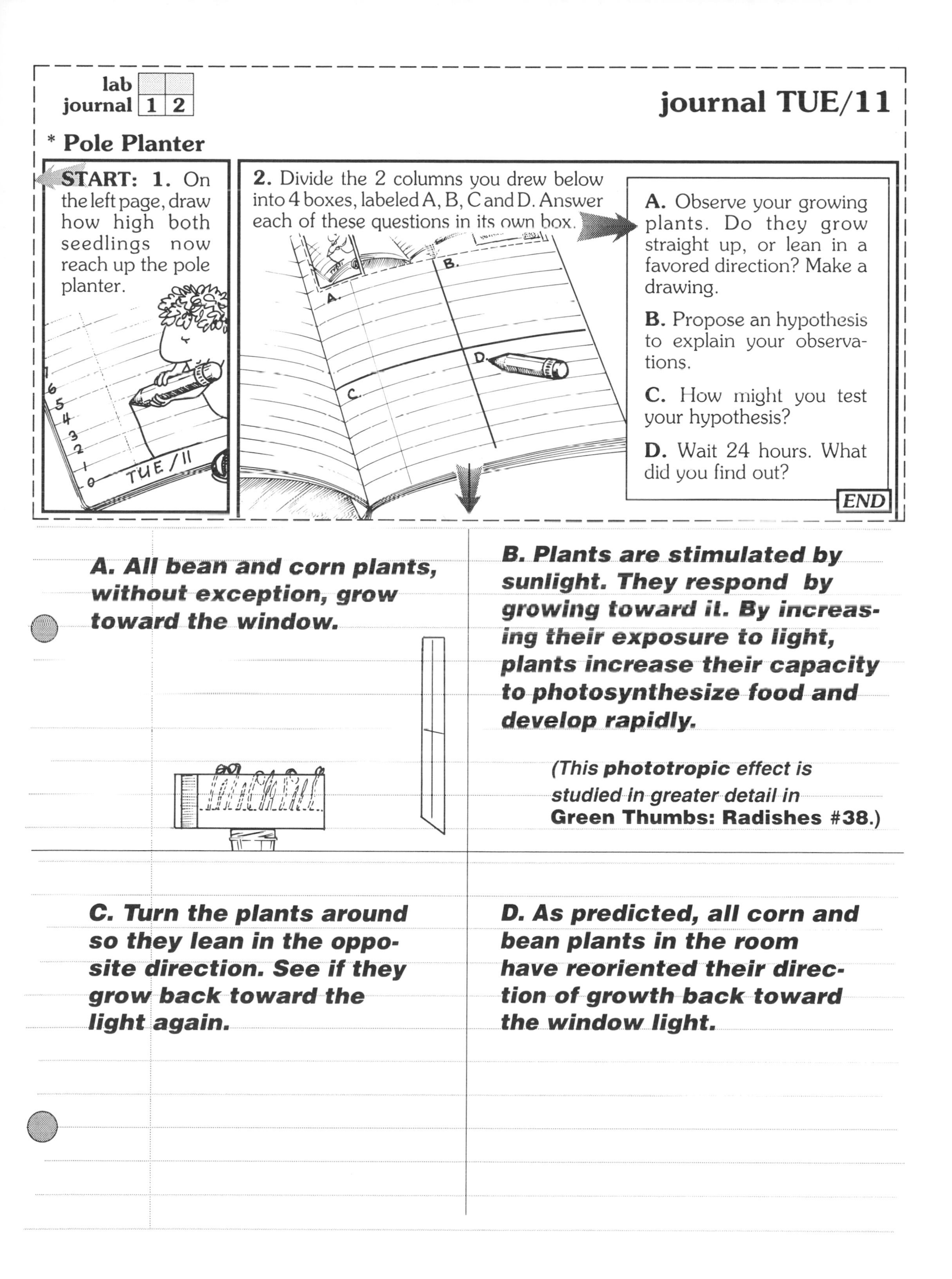
lab journal 1 2
journal TUE/11
* Pole Planter
START: 1. On the left page, draw how high both seedlings now reach up the pole planter.
7 6 5 4 3 2 1 0
TUE/11
2. Divide the 2 columns you drew below into 4 boxes, labeled A, B, C and D. Answer each of these questions in its own box.
A. B. C. D.
A. Observe your growing plants. Do they grow straight up, or lean in a favored direction? Make a drawing.
B. Propose an hypothesis to explain your observations.
C. How might you test your hypothesis?
D. Wait 24 hours. What did you find out?
END
A. All bean and corn plants, without exception, grow toward the window.
B. Plants are stimulated by sunlight. They respond by growing toward it. By increasing their exposure to light, plants increase their capacity to photosynthesize food and develop rapidly.
(This phototropic effect is studied in greater detail in Green Thumbs: Radishes #38.)
C. Turn the plants around so they lean in the opposite direction. See if they grow back toward the light again.
D. As predicted, all corn and bean plants in the room have reoriented their direction of growth back toward the window light.

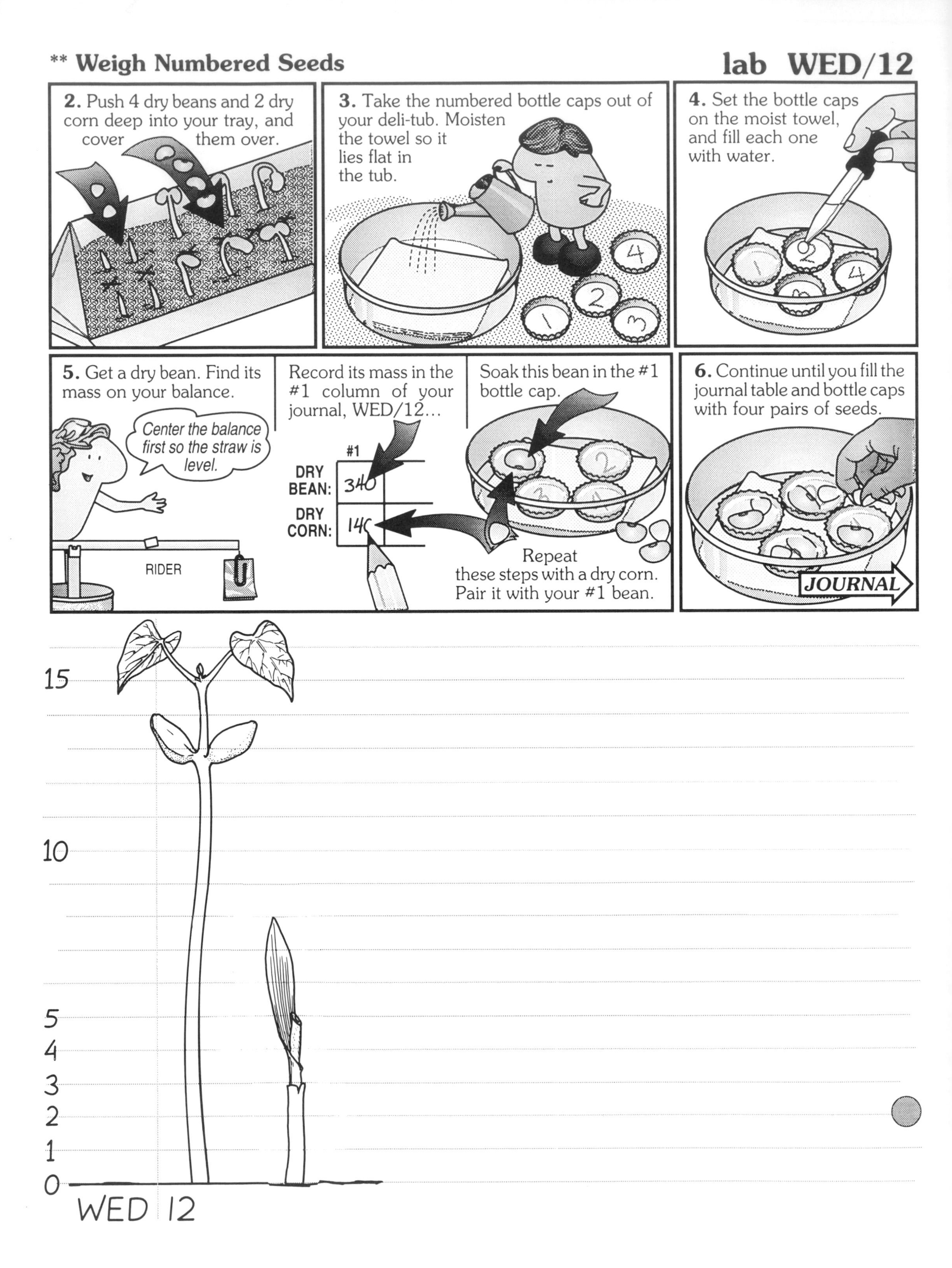
** Weigh Numbered Seeds
lab WED/12
2. Push 4 dry beans and 2 dry corn deep into your tray, and cover them over.
3. Take the numbered bottle caps out of your deli-tub. Moisten the towel so it lies flat in the tub.
4. Set the bottle caps on the moist towel, and fill each one with water.
5. Get a dry bean. Find its mass on your balance.
Center the balance first so the straw is level.
RIDER
Record its mass in the #1 column of your journal, WED/12…
#1
DRY BEAN: 340
DRY CORN: 140
Soak this bean in the #1 bottle cap.
Repeat these steps with a dry corn. Pair it with your #1 bean.
6. Continue until you fill the journal table and bottle caps with four pairs of seeds.
JOURNAL
15
10
5
4
3
2
1
0
WED 12

lab		2	3	4	5	6			
journal	1				5	6	7	8	9

* Pole Planter

START: 1. On the left page, draw each pole plant exactly as you see it — like a snapshot.

LAB

5-6.

Weigh each pair, record, and place in numbered cap.

(all data in milligrams)

	Cap #1	#2	#3	#4	Total	Average
DRY BEAN:	340	410	360	390	1500	375
DRY CORN:	140	150	110	120	520	130

7. Close the lid on your deli tub.

8. Do you think the seeds will gain mass by tomorrow? Explain below.

9. Based on your observations of TUE/4, which seeds will likely gain the most mass?

END

Yes. All the seeds will gain mass because they will all absorb water.

The beans will likely absorb more water than the corn. This is what happened before.

TEACHING NOTES

Remind students to return to question 2D from yesterday, and evaluate their hypotheses.

1. Some students, in their zeal to grow a really tall bean plant, may pull on it to reach higher up the pole. This defeats the purpose of documenting how plants grow naturally. You can discourage this competitive behavior (which could easily damage the stem) by pointing out that plant length is not a good indicator of plant health. Beans that don't receive enough sunlight, for example, grow very tall and spindly. Remind students to draw each day's growth exactly as they find it, without striking an unnatural pose.

It may be helpful to demonstrate how to draw the bean and corn plants life-size by sketching an example on your blackboard. Begin by drawing and numbering ruled lines to represent notebook paper. Then transfer points of reference from an actual bean pole plant to your blackboard, fixing the growing tip, leaves and cotyledons at the correct heights. Finally, fill in the interconnecting stems. Advise students to turn the jar to the same perspective each new day. In this manner, they can make tracings of previous development, then add on what's new.

If the bean leans too far from its support pole, add another masking-tape tie. Always turn the jar so the support pole stands between the bean plant and its primary light source. This insures that the plant will grow toward its pole as its leaves seek maximum daylight.

Challenge your more capable students to capture fine details. They can draw leaf veins, for example, on just one leaf or all the leaves. If you provide colored pencils, students can record the many different shades of yellow and green, and document how they change over time.

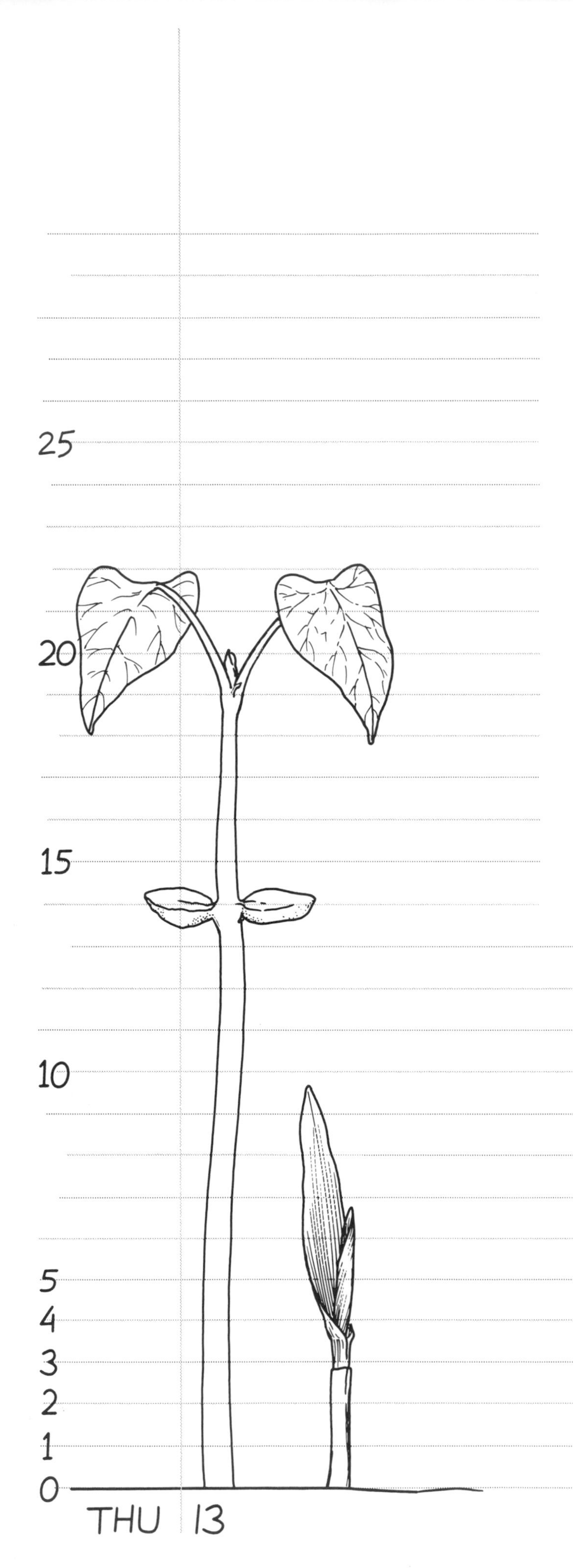

TEACHING NOTES

1. The bean will soon grow higher than the journal page. When this happens, advise students to split their bean drawing into two parts: the lower bean to the left (from ground level to above the cotyledons); the upper bean to the right (from above the cotyledons to the top of the plant). Number the lines on each side of the page accordingly, as first shown in TUE/18.

2. Each numbered bottle cap must be emptied of water and returned to its sealed deli tub containing only its moist seed pair. The moist atmosphere sealed in the tub will prevent the seeds from drying out. Having already absorbed water, these seeds now require oxygen to metabolize stored food reserves and begin sprouting. If seeds are kept in water, they will sprout less vigorously or begin to rot.

Each pair of seeds is numbered by association with its particular bottle cap. Caution students not to confuse these numbers by mixing seeds and bottle caps. This is easily prevented it students weigh seed pairs separately, as directed. Remove, empty, weigh and return #1 first. Then repeat these operations for the seeds in cap #2. Then do cap #3. And finally, #4.

Calculate the water absorbed by subtracting the seed's smaller dry mass from its larger wet mass. There is plenty of room to show all calculation on the left journal page. Discourage those who erase their old figures with every new calculation, as if they were trying to hide arithmetic errors. Just the opposite is desirable. Let the figures stand; if there is an error, it will be easier to trace and fix.

3. The seeds *must* be *sealed* inside the moist deli tub. Otherwise they will cool by evaporation and dry out.

lab journal | 1 | 2 | 3 | 4 | 5

journal THU/13

* Pole Planter

START: **1.** Draw each pole plant exactly as you see it. Your sketch should always show the height of leaves and cotyledons.

LEAF NODE

COTYLEDON

2. Open your deli tub. Put the seeds from Bottle Cap #1 on a dry paper towel and empty out the water… Weigh each seed to complete this table. Return each seed pair to its numbered cap (without water) before weighing the next.)

3. Seal your deli tub. Don't open it again until MON/17.

(all data in milligrams)

	Cap #1	#2	#3	#4	Total	Average
SOAKED BEAN:	690	810	710	780	2990	748
DRY BEAN:	340	410	360	390	1500	375
Water Absorbed:	350	400	350	390	1490	373
SOAKED CORN:	180	200	150	170	700	175
DRY CORN:	140	150	110	120	520	130
Water Absorbed:	40	50	40	50	180	45

4. Use data from your table to compare how beans and corn absorb water. Was your prediction from WED/12 a good one?

My data suggests…

5. How many beans are needed to absorb a glassful of water? (1 gram = 1,000 mg)

250 g

one glass full!

END

(Answers will vary. Encourage as much math analysis as possible.)

GOOD ANSWER: The beans absorbed an average of 373 mg of water per seed, while the corn only absorbed 45 mg of water per seed. As predicted yesterday, the beans gained far more mass than the corn.

BETTER ANSWER: The beans absorbed an average of 373 mg of water. This represents nearly a 100% increase in dry mass:

$$\frac{373}{375} \times 100 = 99\% \text{ increase}$$

The corn absorbed, on average, only 35% of its dry mass in water:

$$\frac{45}{130} \times 100 = 35\% \text{ increase}$$

As predicted yesterday, the beans gained far more mass than the corn.

(If this problem is too difficult for younger students, try doing it on the blackboard as a class exercise.)

According to the illustration, a glass of water has a mass of 250 g, or 250,000 mg. If an average bean absorbs 373 mg, then the total number of beans that will absorb a glassful of water can be found by division:

$$\frac{250{,}000 \text{ mg}}{\text{glassful}} \times \frac{1 \text{ bean}}{373 \text{ mg}} = \frac{670 \text{ beans}}{\text{glassful}}$$

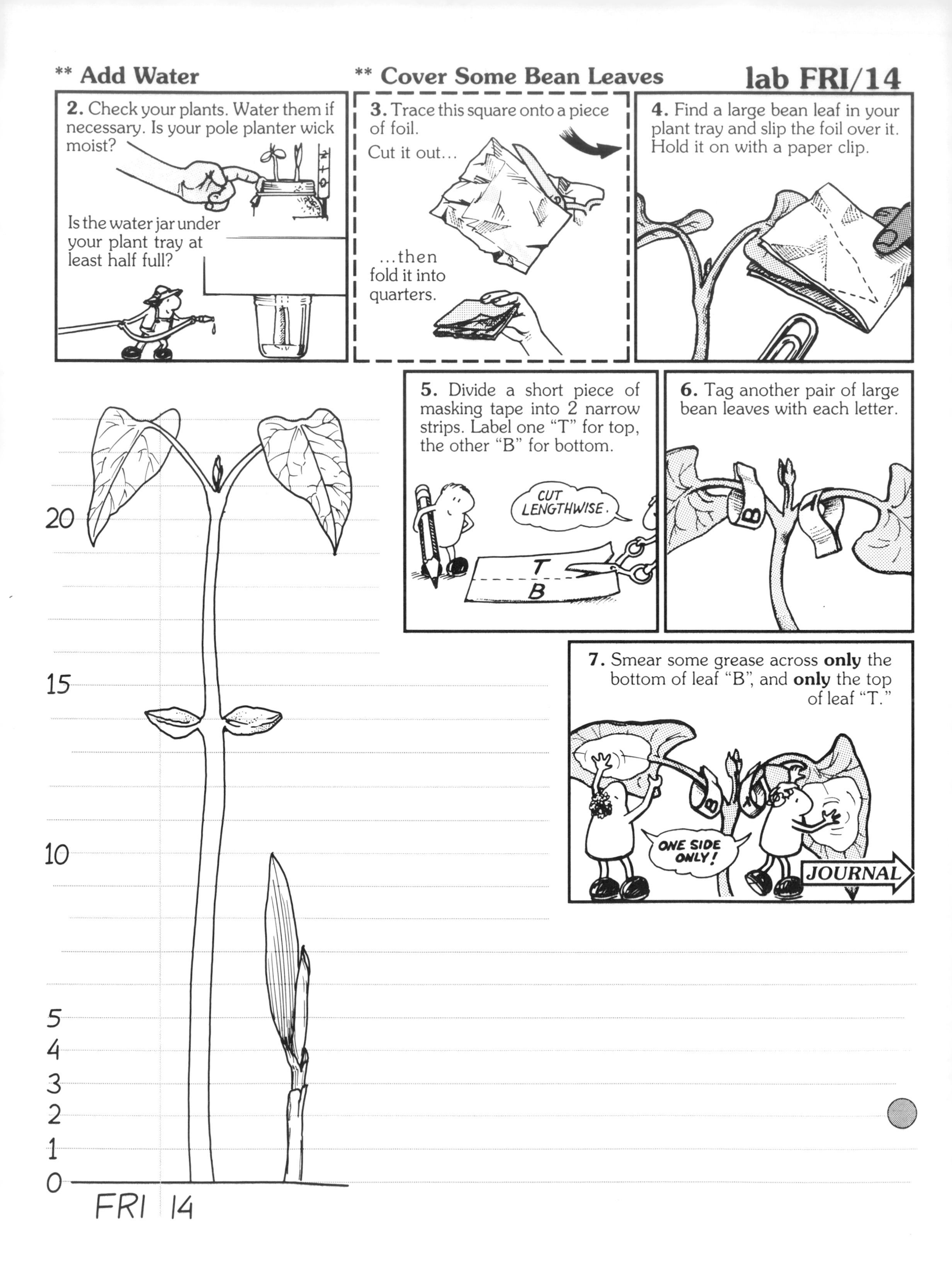

** Add Water
** Cover Some Bean Leaves
lab FRI/14
2. Check your plants. Water them if necessary. Is your pole planter wick moist?
Is the water jar under your plant tray at least half full?
3. Trace this square onto a piece of foil.
Cut it out…
…then fold it into quarters.
4. Find a large bean leaf in your plant tray and slip the foil over it. Hold it on with a paper clip.
5. Divide a short piece of masking tape into 2 narrow strips. Label one "T" for top, the other "B" for bottom.
CUT LENGTHWISE.
T
B
6. Tag another pair of large bean leaves with each letter.
B
T
7. Smear some grease across only the bottom of leaf "B", and only the top of leaf "T."
B
T
ONE SIDE ONLY!
JOURNAL
20
15
10
5
4
3
2
1
0
FRI 14

lab		2	3	4	5	6	7		
journal	1							8	9

journal FRI/14

* Pole Planter

START: 1. Accurately draw on the left page how both plants have gained height. LAB

8. Leaves turn green in sunlight as they use the sun's energy to photosynthesize their own food.

Green!

Predict the color of your leaf after being covered for a long time with foil. Explain your reasons.

The part of the leaf covered by foil won't receive sunlight. Thus it can't photosynthesize, and will lose its green color. It will turn more and more yellow and eventually die.

9. Leaves breathe air through tiny openings (stomata) in their leaves. Stomata are too small to see without a microscope. If grease plugs these holes, the leaves can't breathe.

CO_2

O_2

How might this experiment tell us where the stomata are located?

END

That part of the leaf with plugged stomata won't be able to breathe or carry on photosynthesis. Over time, it will likely turn yellow and die.

If "B" yellows but "T" remains green, then stomata must be located only on the bottom of the leaf. If the opposite happens, then stomata must be located only on the top.

If no part of either leaf dies, perhaps the experiment hasn't run long enough, or stomata are located on both sides. If both parts on each leaf die, perhaps the petroleum jelly is toxic to the plant.

TEACHING NOTES

2. Check for standing water in the pole planters. Students tend to over-water. Gently invert the planters over a sink to drain away any excess. Vermiculite will not spill out of the jar.

To avoid congestion or excessive traffic to and from a single water source, provide two-liter watering bottles at various locations around your room.

7. Be sure your class understands the purpose of this experiment *before* applying grease to their fingertips. Common mistakes are using too much grease, and applying it to both sides of the leaf.

Additional Materials

- ☐ Two-liter watering bottles (optional).
- ☐ A jar of petroleum jelly.

** Tie a Bean and Corn Sprout

2. Open your deli-tub and remove your most advanced bean sprout. *Gently* tie a thread on its hypocotyl.

DOUBLE KNOT

3. Tie another thread around your most advanced corn sprout, just behind its coleoptile.

Record both seed numbers on scratch paper.

Trim one end of each thread short: leave one end as long as your hand.

4. Write the bottle cap number of each sprout in your journal table…

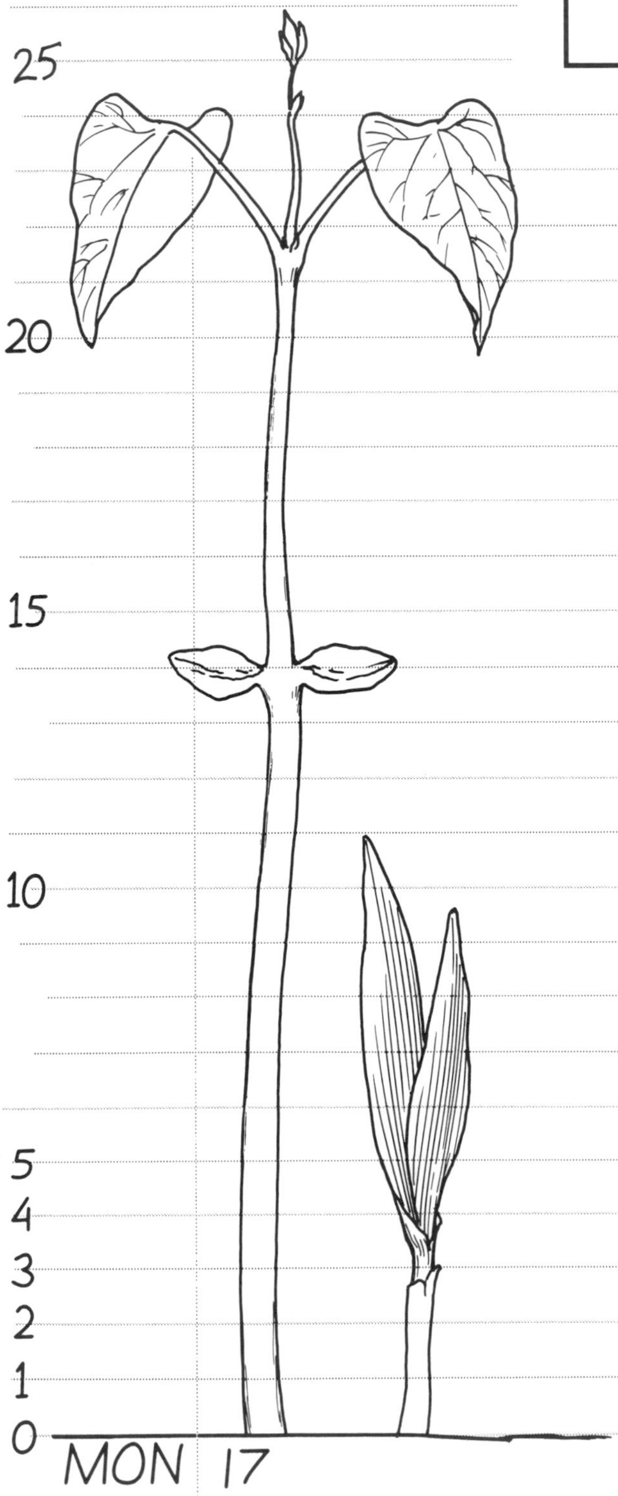

(all data in milligrams)		Wed/12	Thu/13
BEAN:	# 4		
CORN:	#		

Discard your remaining sprouts.

m-m-m-m-m— Snacks?!

free sprouts

JOURNAL

25

20

15

10

5
4
3
2
1
0

MON 17

lab		2	3	4			
journal	1			4	5	6	7

journal MON/17

* Pole Planter

START: **1.** Draw accurately how both plants have grown.

LAB

4. Complete this mass table. Be sure to center your balance first...

(all data in milligrams)		DRY Wed/12	SOAKED Thu/13	SPROUTED Mon/17
BEAN:	# 4	390	780	1090
CORN:	# 2	150	200	230

5. Clean out your deli tub, and line it with 3 clean, moist towels. Keep both weighed sprouts inside with the lid closed.

6. How much *more* mass did each seed gain after the 24-hour soak of THU/13?

7. What *other* way can seedlings gain mass besides absorbing water? END

The bean gained 310 mg of additional mass, while the corn gained 30 mg.

BEAN: 1090 − 780 = 310 ***CORN:*** 230 − 200 = 30

The seedlings can turn green and begin photo-synthesizing their own food.

TEACHING NOTES

2-4. Once students select their most advanced bean and corn sprout, they should immediately record each number on scratch paper, as directed. This avoids any potential mix up before students enter these numbers in their journal data tables (step 4).

Tying thread to each sprout provides a convenient tie-in to the balance for daily weighings. For now the sprouts may be small enough to deposit directly in the foil weighing cup. But as the seedlings grow larger, students will need to tie them onto the foil cup, then suspend them over the table edge to determine mass.

Should the mass of the thread be subtracted from the mass of each sprout? For greatest accuracy, yes. (Praise students who even think to ask such a question.) But since this thread mass is quite small compared to the mass of each sprout we can reasonably ignore it here.

2-3. These steps require a delicate hand. Students who inadvertently damage their sprouts can try again with new ones (and record new numbers on scratch paper). Or you might provide direct assistance. Out of a total of 8 sprouts, only 1 bean and 1 corn sprout need to be successfully tied (and their numbers accurately recorded).

The bean's hypocotyl is easiest to tie with thread. But care must be taken not to pull the loop too tightly and cut into soft tissue. To underscore this point, ask students how it might feel to have a thread loop pulled too tightly around one of their fingers.

If the corn's coleoptile protrudes, as illustrated, it will prevent the thread loop from slipping off the tapered end of the seed when it is drawn closed and knotted. If this structure has not yet emerged, tie the thread to the base of the tap root.

All untied survivors can be either discarded, taken home, or added to the backup garden you may have set up in step 8 of MON/10.

5. Set the two tied and weighed sprouts directly on the fresh, moist towels. Numbered bottle caps are no longer needed, since the numbers are already recorded for this pair.

Why are there now three towels in the tub? For now, they hold enough water to insure that the sprouts don't dry out. Later, two of the towels will be used to cover the roots when the sprouts grow too large to be housed inside the sealed container.

* Compare Leaves

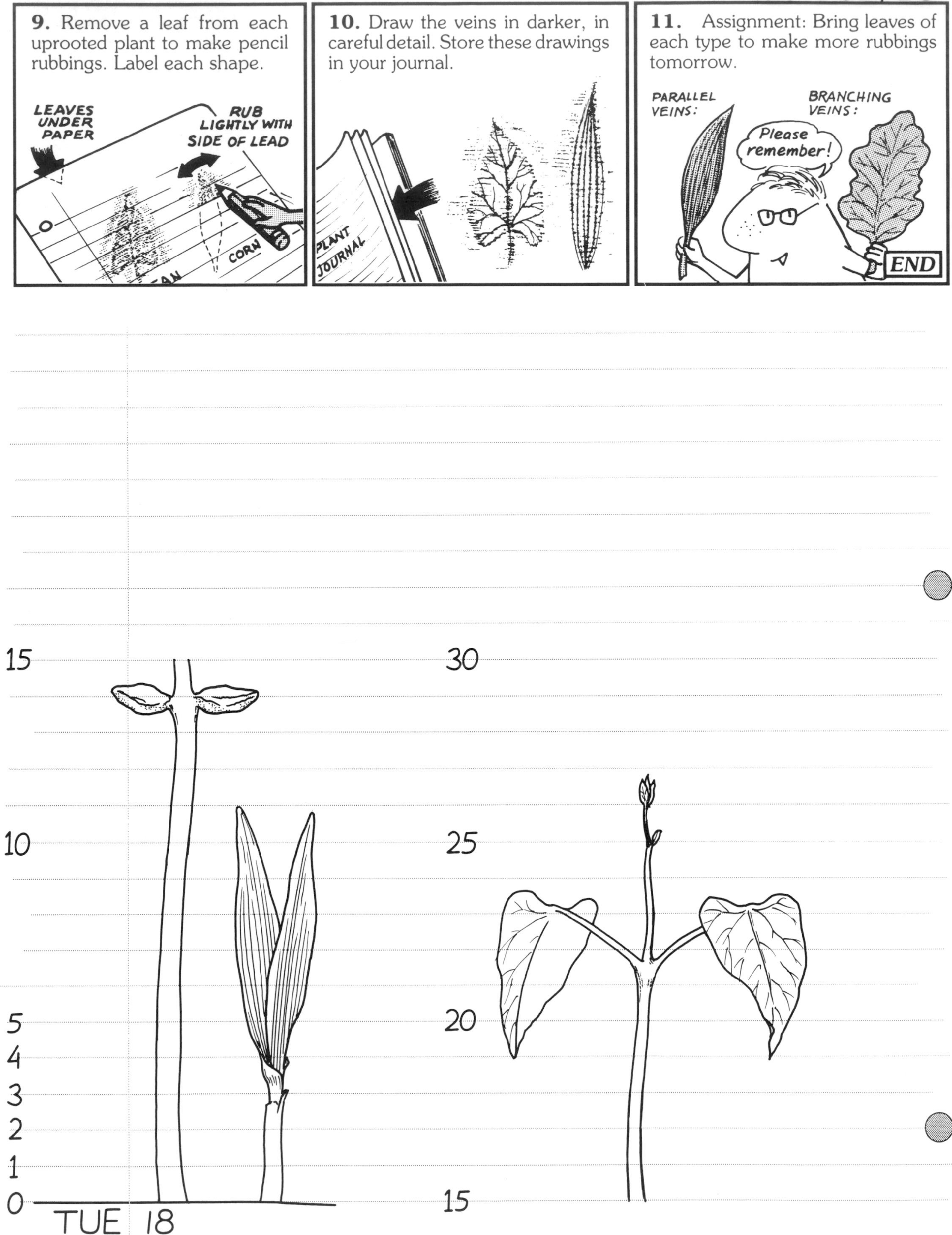

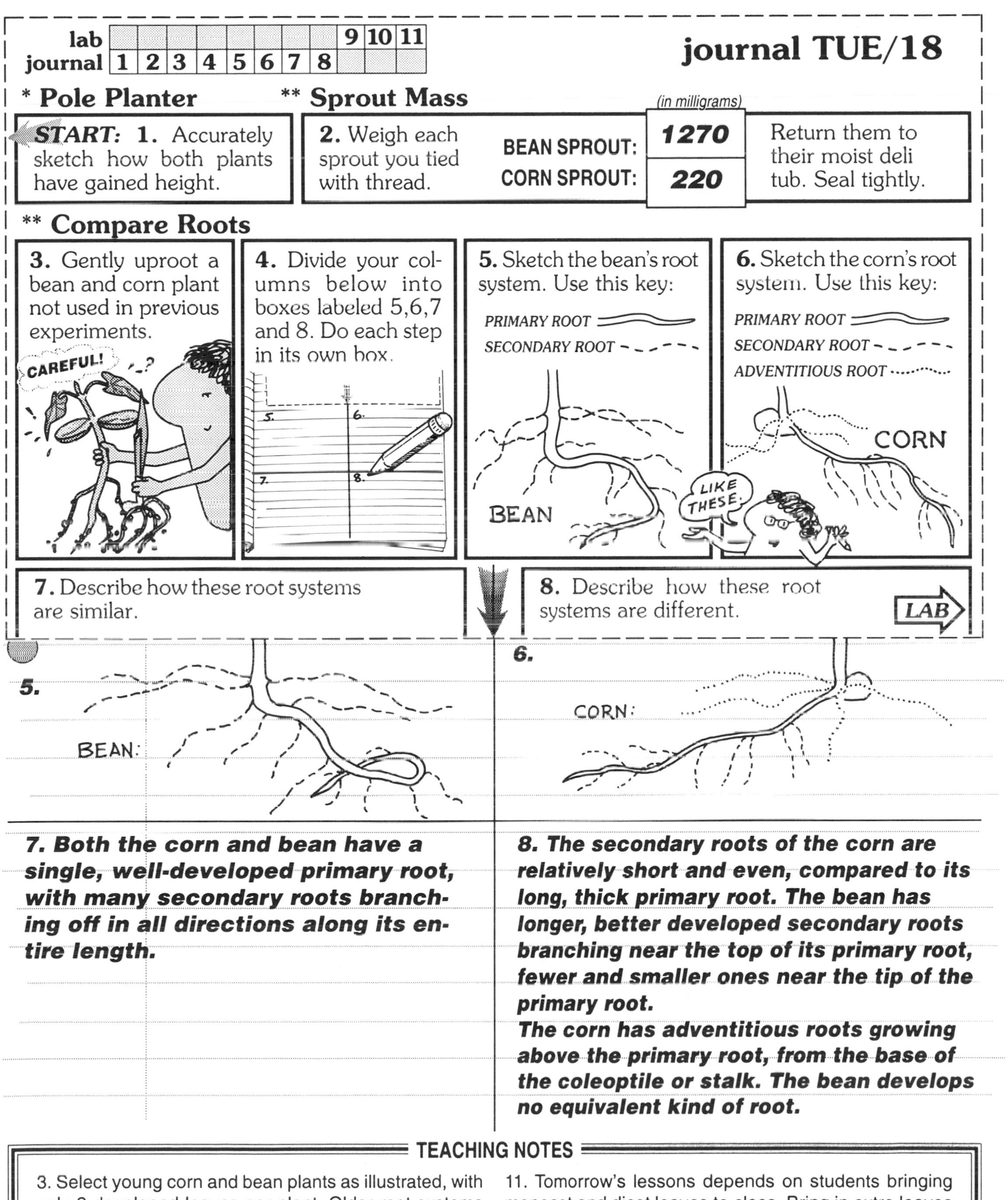

lab 9 10 11
journal 1 2 3 4 5 6 7 8

journal TUE/18

* **Pole Planter**

START: **1.** Accurately sketch how both plants have gained height.

** **Sprout Mass**

2. Weigh each sprout you tied with thread.

	(in milligrams)
BEAN SPROUT:	**1270**
CORN SPROUT:	**220**

Return them to their moist deli tub. Seal tightly.

** **Compare Roots**

3. Gently uproot a bean and corn plant not used in previous experiments.

4. Divide your columns below into boxes labeled 5,6,7 and 8. Do each step in its own box.

5. Sketch the bean's root system. Use this key:

6. Sketch the corn's root system. Use this key:

7. Describe how these root systems are similar.

8. Describe how these root systems are different. LAB

7. Both the corn and bean have a single, well-developed primary root, with many secondary roots branching off in all directions along its entire length.

8. The secondary roots of the corn are relatively short and even, compared to its long, thick primary root. The bean has longer, better developed secondary roots branching near the top of its primary root, fewer and smaller ones near the tip of the primary root.
The corn has adventitious roots growing above the primary root, from the base of the coleoptile or stalk. The bean develops no equivalent kind of root.

TEACHING NOTES

3. Select young corn and bean plants as illustrated, with only 2 developed leaves per plant. Older root systems are too complex to easily draw.

11. Tomorrow's lessons depends on students bringing monocot and dicot leaves to class. Bring in extra leaves yourself to share with those who forget. Write this assignment on your blackboard for emphasis.

TEACHING NOTES

2. These bean and corn sprouts will grow quickly over the next several days, and will soon outgrow the confined space in their sealed deli tub. If this has happened already, follow the instructions below to adapt your sprouts to an unsealed growing environment. If they are still small, students might wait until FRI/21 to make this change. Don't wait any longer, since your sprouts will almost certainly require more growing room over the weekend.

RAISING THE ROOF:

(a) Remove all sprouts and towels from the deli tub. Gently unfold one of the wet towels, then loosely crumple it to create "root pockets" in the bottom of the tub.

(b) "Plant" each sprout gently in a fold of this towel, draping one of the other wet towels over each root system like a blanket.

(c) Fill the tub about 1/3 full with water. Refill as necessary every 2 or 3 days so the roots always remain moist.

3-4. All flowering plants divide into 2 broad groups — dicots and monocots. There are roughly 200,000 species of dicots and 50,000 species of monocots. If time allows, go outside and classify green plants growing in your school yard. All vascular plants except ferns and evergreens (these are non flowering) can be included. Bring back a collection of leaves and seeds for a bulletin board display.

Additional Materials

☐ Leaf samples, both monocot and dicot.

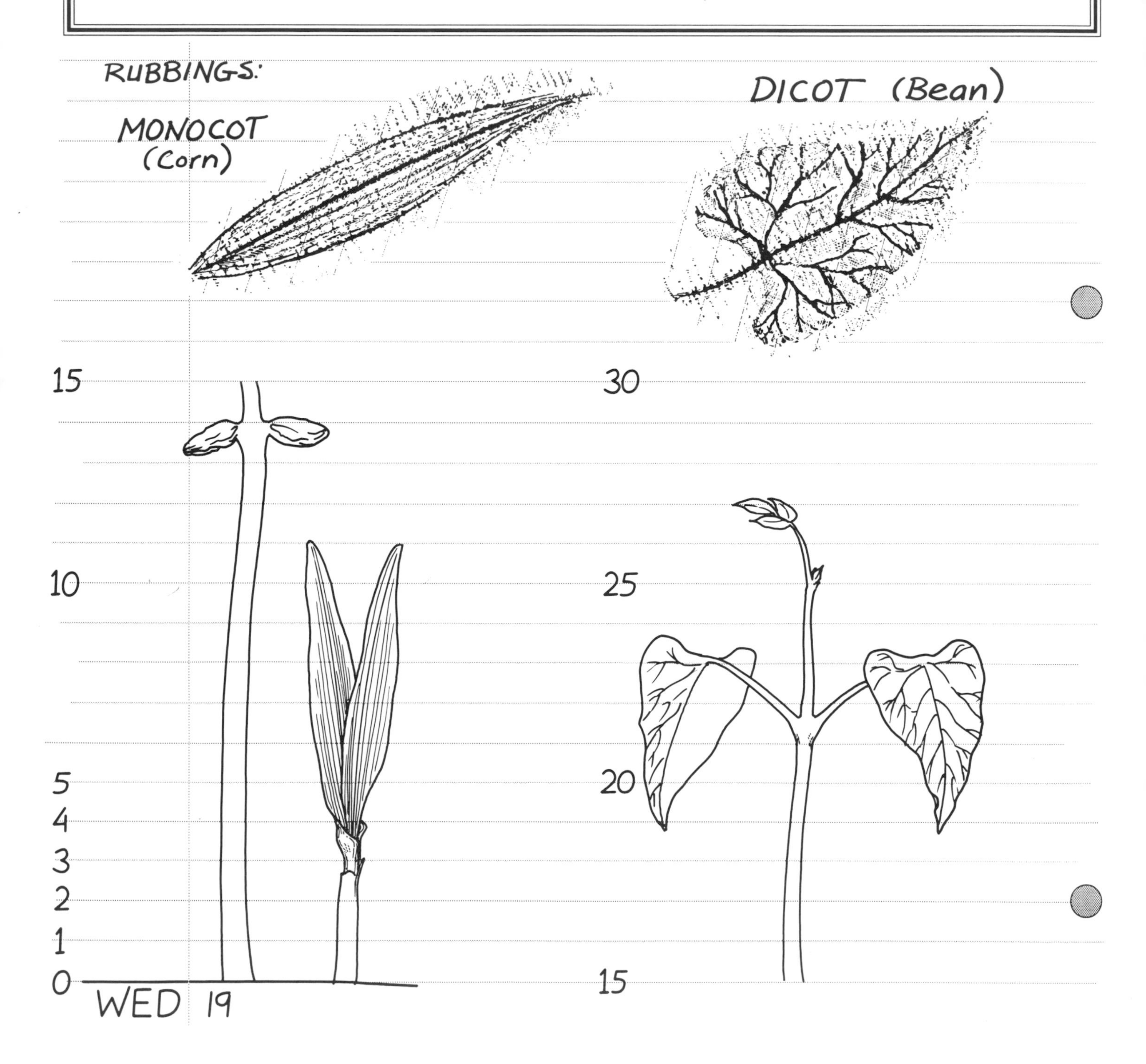

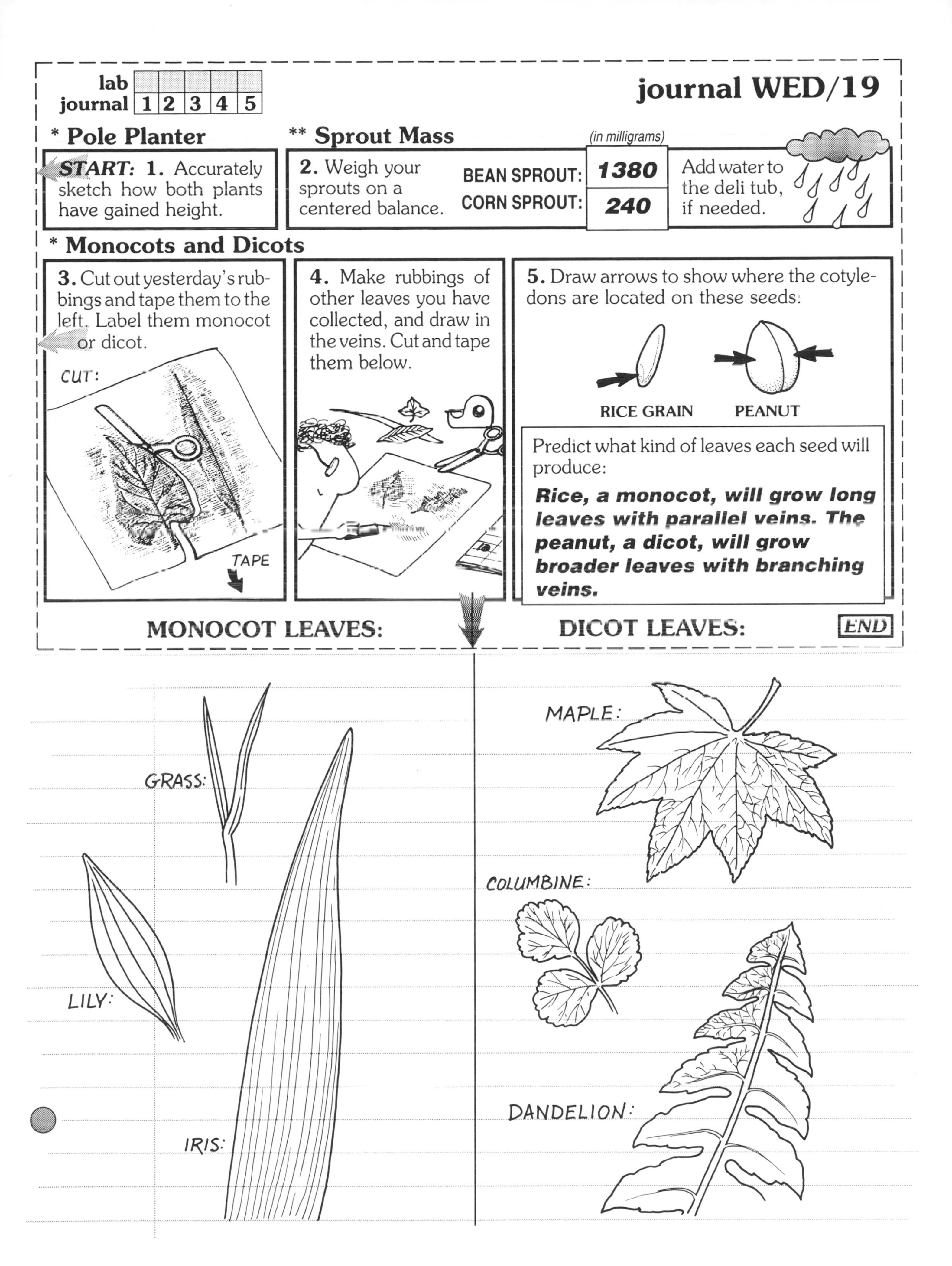
lab
journal 1 2 3 4 5
journal WED/19
* Pole Planter
START: 1. Accurately sketch how both plants have gained height.
** Sprout Mass
2. Weigh your sprouts on a centered balance.
(in milligrams)
BEAN SPROUT: 1380
CORN SPROUT: 240
Add water to the deli tub, if needed.
* Monocots and Dicots
3. Cut out yesterday's rubbings and tape them to the left. Label them monocot or dicot.
CUT:
TAPE
4. Make rubbings of other leaves you have collected, and draw in the veins. Cut and tape them below.
5. Draw arrows to show where the cotyledons are located on these seeds:
RICE GRAIN
PEANUT
Predict what kind of leaves each seed will produce:
Rice, a monocot, will grow long leaves with parallel veins. The peanut, a dicot, will grow broader leaves with branching veins.
MONOCOT LEAVES:
DICOT LEAVES:
END
GRASS:
LILY:
IRIS:
MAPLE:
COLUMBINE:
DANDELION:

** **Study Cotyledons**

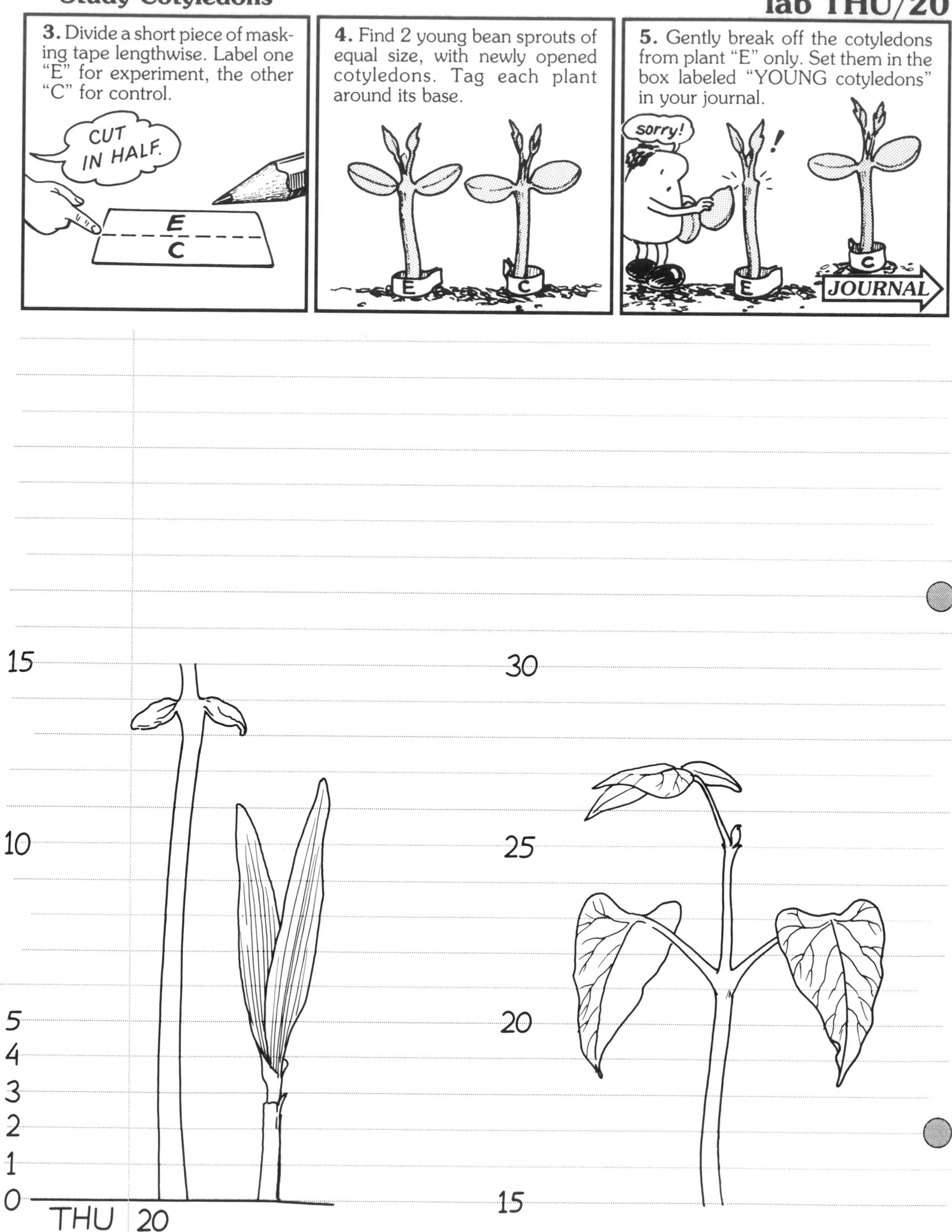

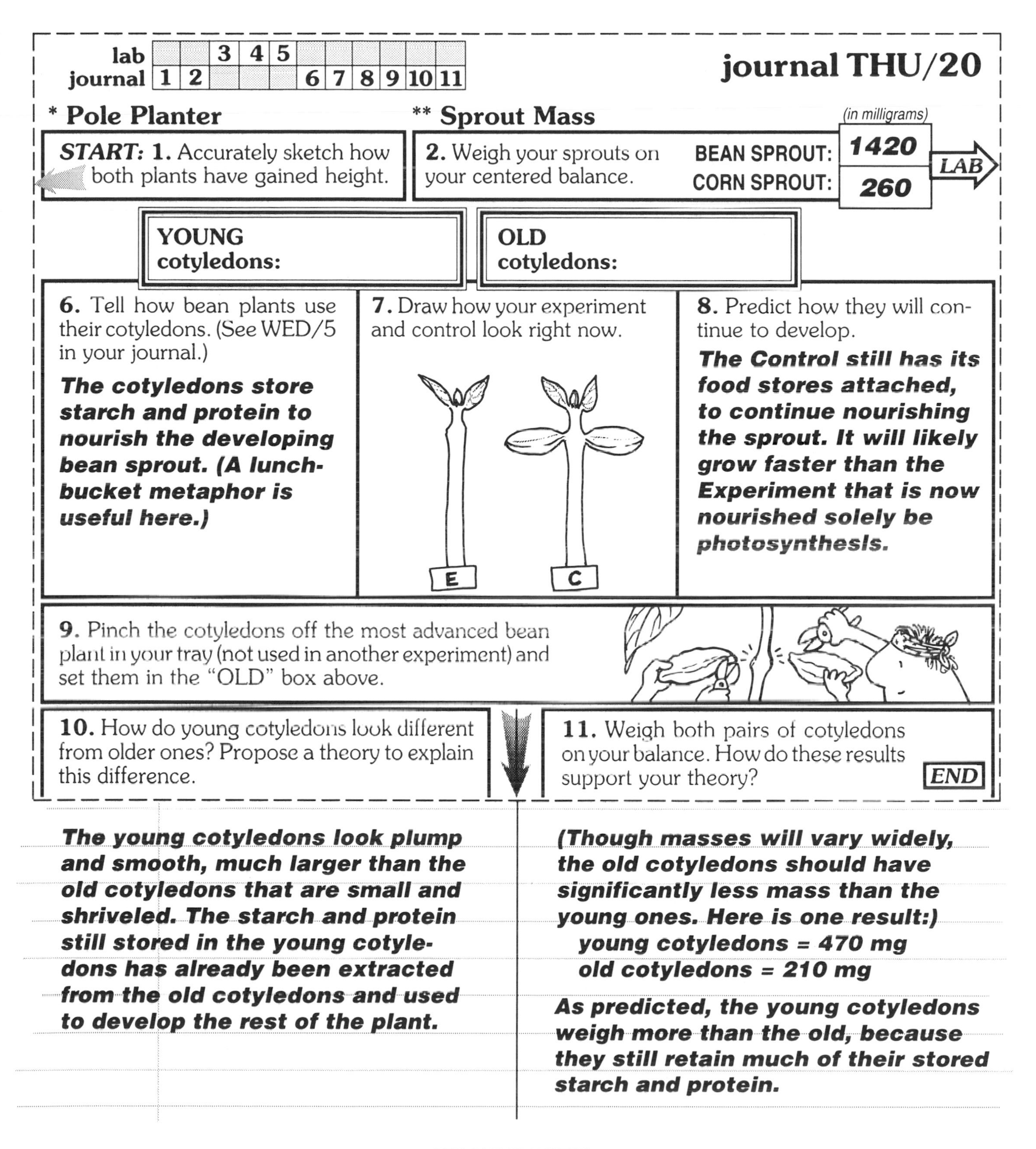

lab			3	4	5						
journal	1	2				6	7	8	9	10	11

journal THU/20

* **Pole Planter**

** **Sprout Mass**

START: 1. Accurately sketch how both plants have gained height.

2. Weigh your sprouts on your centered balance.

(in milligrams)

BEAN SPROUT:	***1420***
CORN SPROUT:	***260***

LAB

YOUNG cotyledons:

OLD cotyledons:

6. Tell how bean plants use their cotyledons. (See WED/5 in your journal.)

The cotyledons store starch and protein to nourish the developing bean sprout. (A lunch-bucket metaphor is useful here.)

7. Draw how your experiment and control look right now.

8. Predict how they will continue to develop.

The Control still has its food stores attached, to continue nourishing the sprout. It will likely grow faster than the Experiment that is now nourished solely be photosynthesis.

9. Pinch the cotyledons off the most advanced bean plant in your tray (not used in another experiment) and set them in the "OLD" box above.

10. How do young cotyledons look different from older ones? Propose a theory to explain this difference.

11. Weigh both pairs of cotyledons on your balance. How do these results support your theory?

END

The young cotyledons look plump and smooth, much larger than the old cotyledons that are small and shriveled. The starch and protein still stored in the young cotyledons has already been extracted from the old cotyledons and used to develop the rest of the plant.

(Though masses will vary widely, the old cotyledons should have significantly less mass than the young ones. Here is one result:)

young cotyledons = 470 mg

old cotyledons = 210 mg

As predicted, the young cotyledons weigh more than the old, because they still retain much of their stored starch and protein.

TEACHING NOTES

4-5. Select experiment and control plants that are as young as possible. These will display a much greater difference in growth rates than older plants.

9. Students should leave the cotyledons on their pole planter beans intact. As students continue to sketch their daily development, they'll soon discover that the cotyledons drop off naturally, without outside assistance.

TEACHING NOTES

2-3. If your students haven't yet done so, they may need to leave their deli tubs uncovered so the sprouts have growing room over the weekend. Review the teaching notes on WED/19 for important instructions on keeping roots moist.

4. Conserve paper! Urge students to share.

8. To place the next dot, 1 line width away, line up either edge of the paper with the previous dot. Re-ink the pinhead after each dot.

9. To stimulate speculation about how plants grow, draw a thin, straight "stem," marked at equal intervals, on your blackboard. Label it "before." Then draw possible growth outcomes; label these "after." For each case, discuss where new cells formed, where old cells enlarged.

(a) New stem cells formed at the top. BEFORE AFTER

(b) All old stem cells grew a little longer. BEFORE AFTER

(c) New stem cells formed at the bottom, pushing up older growth. BEFORE AFTER

10. Select rapidly growing plants. A good choice for the corn is one with a newly emerging second leaf. Find a bean with 2 leaves fully deployed and a middle shoot pushing up perhaps a centimeter.

Additional Materials

☐ Food coloring, and a small lid to use as a palette.

** Study Plant Growth — lab FRI/21

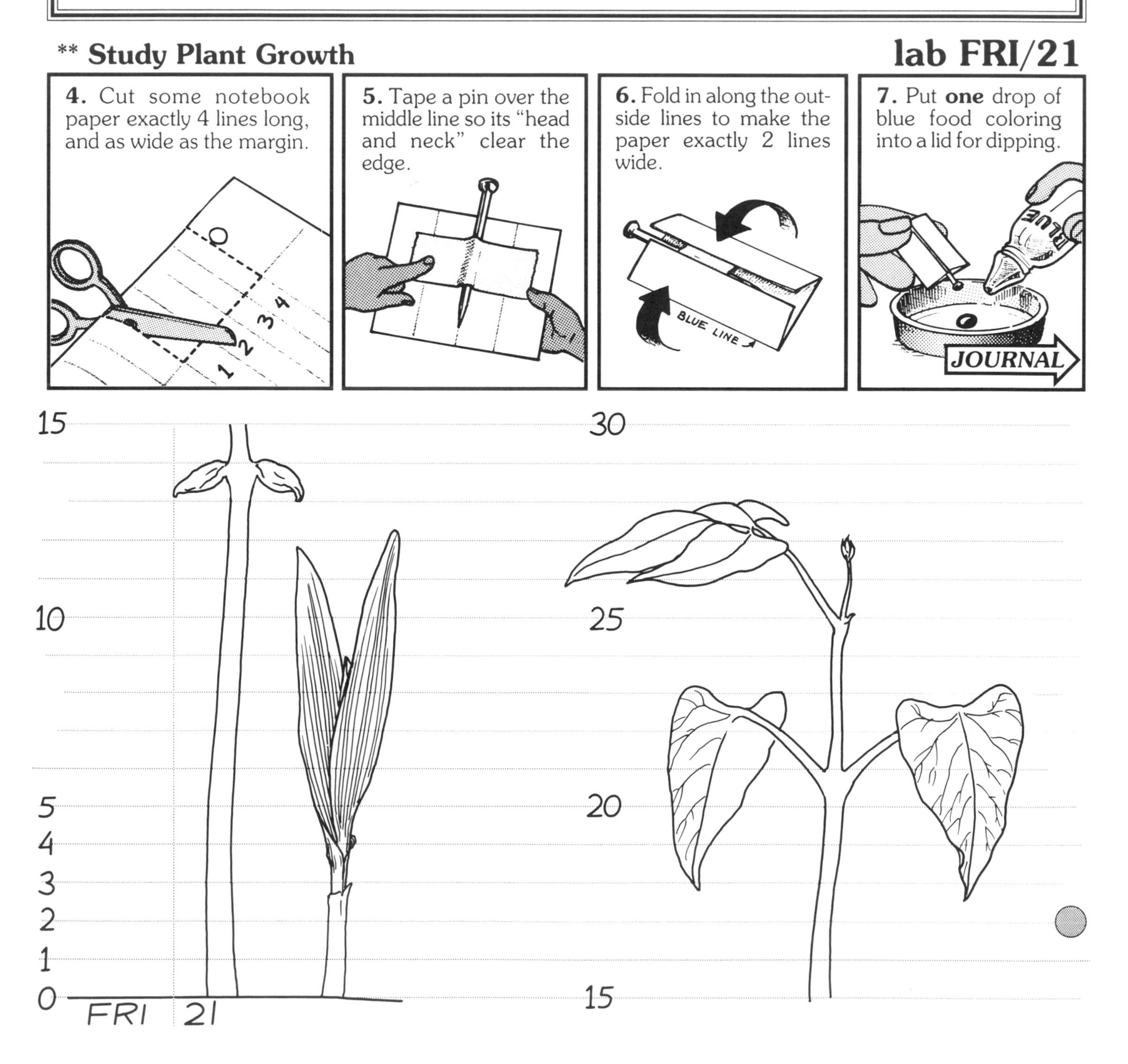

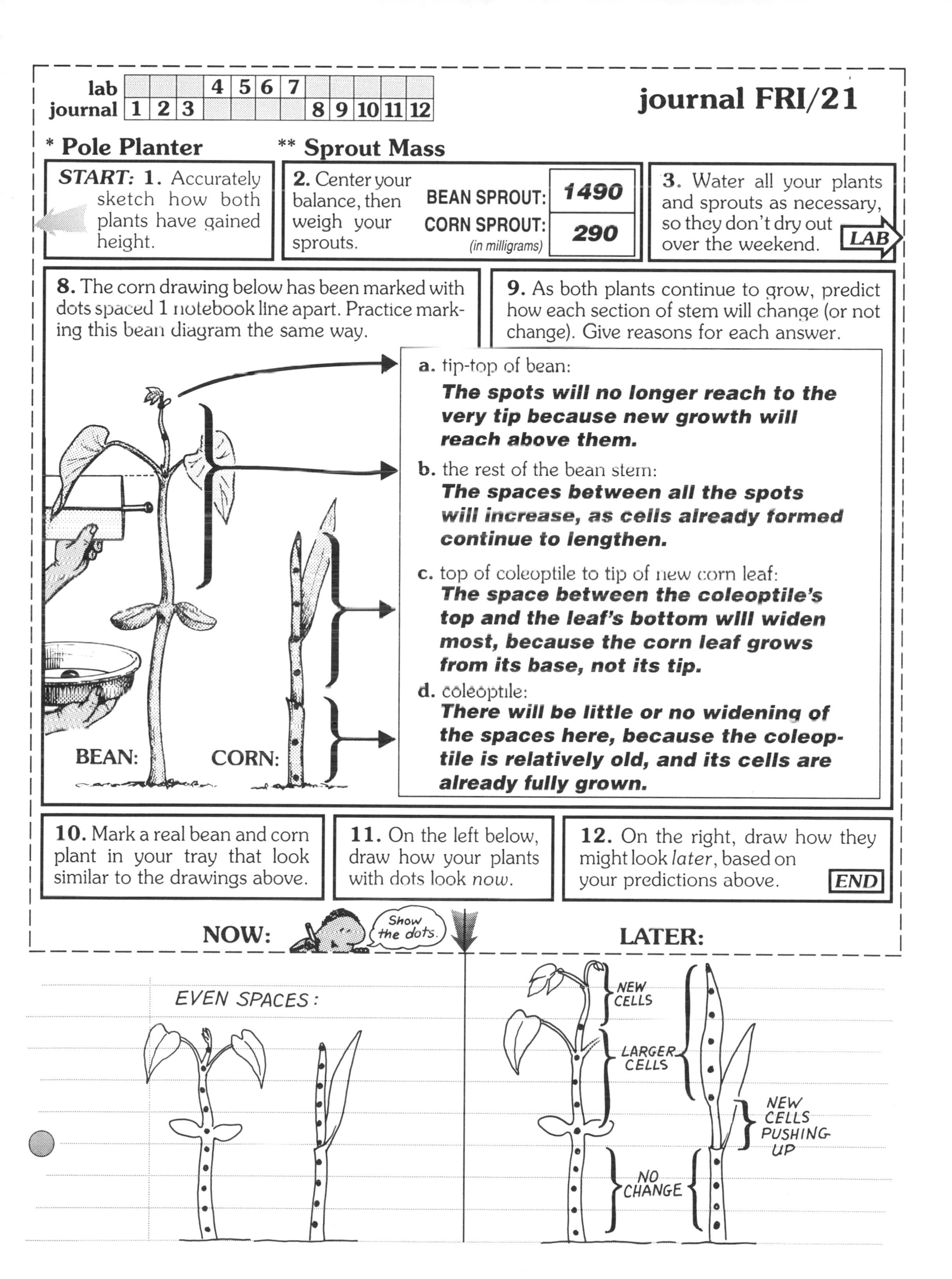

lab				4	5	6	7					
journal	1	2	3					8	9	10	11	12

journal FRI/21

* Pole Planter

** Sprout Mass

START: 1. Accurately sketch how both plants have gained height.

2. Center your balance, then weigh your sprouts.

BEAN SPROUT:	*1490*
CORN SPROUT: (in milligrams)	*290*

3. Water all your plants and sprouts as necessary, so they don't dry out over the weekend. LAB

8. The corn drawing below has been marked with dots spaced 1 notebook line apart. Practice marking this bean diagram the same way.

9. As both plants continue to grow, predict how each section of stem will change (or not change). Give reasons for each answer.

a. tip-top of bean:
The spots will no longer reach to the very tip because new growth will reach above them.

b. the rest of the bean stem:
The spaces between all the spots will increase, as cells already formed continue to lengthen.

c. top of coleoptile to tip of new corn leaf:
The space between the coleoptile's top and the leaf's bottom will widen most, because the corn leaf grows from its base, not its tip.

d. coleoptile:
There will be little or no widening of the spaces here, because the coleoptile is relatively old, and its cells are already fully grown.

10. Mark a real bean and corn plant in your tray that look similar to the drawings above.

11. On the left below, draw how your plants with dots look *now*.

12. On the right, draw how they might look *later*, based on your predictions above. END

TEACHING NOTES

4. This stretching rubber band successfully models plant cell growth in the length dimension only. Real plant cells also increase in diameter, unlike this rubber band which actually narrows.

5. Old cells don't grow indefinitely. To sustain growth, new cells also must also form, through cell division. In this model they are "dividing" as the rubber band slides through the thumb and forefinger of the right hand.

8. If your growing conditions are slow, the triplet growth patterns noted in this model answer may not yet be apparent.

Additional Materials

- ☐ A thick rubber band.

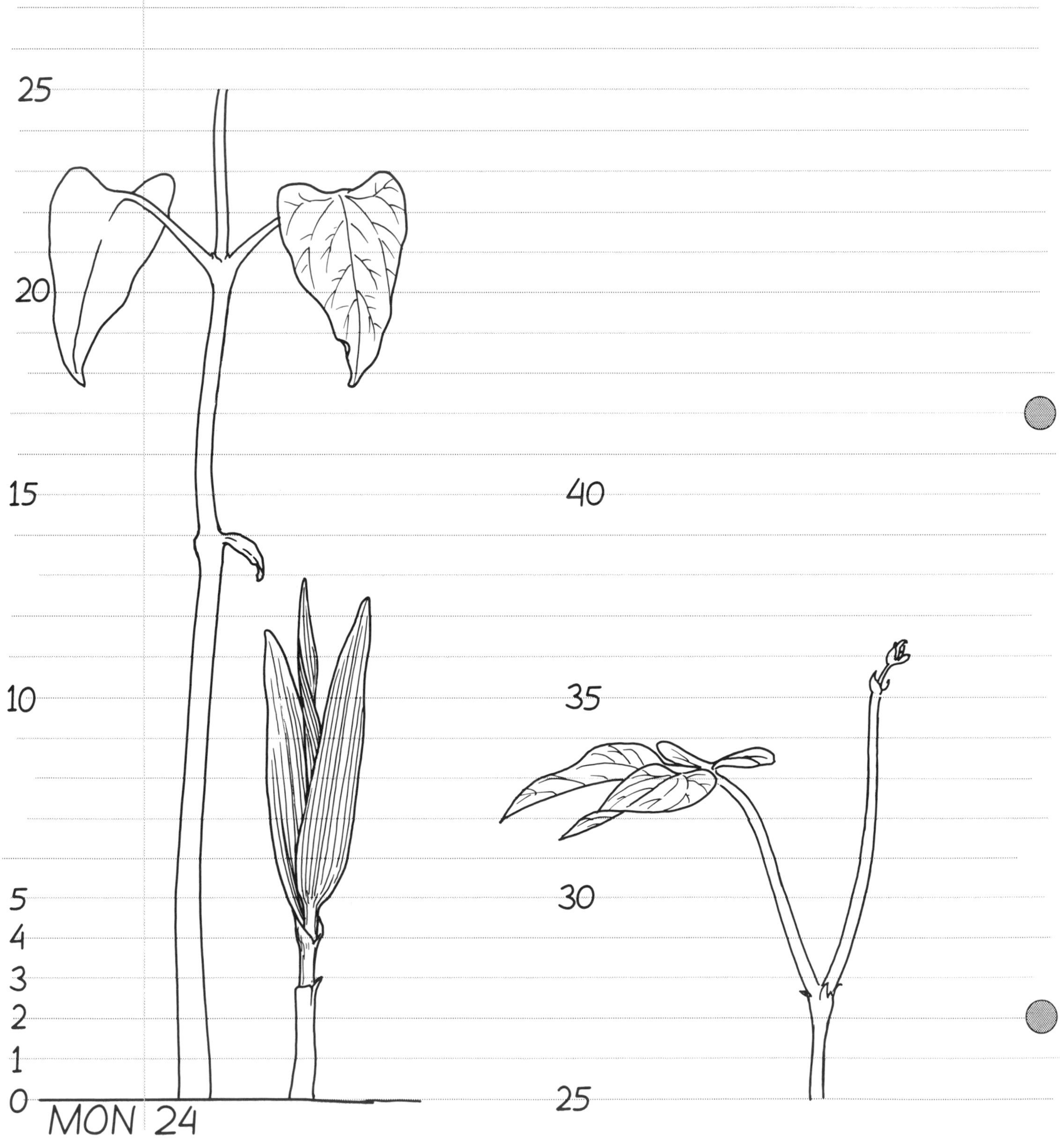

25
20
15
10
5
4
3
2
1
0
MON 24

40
35
30
25

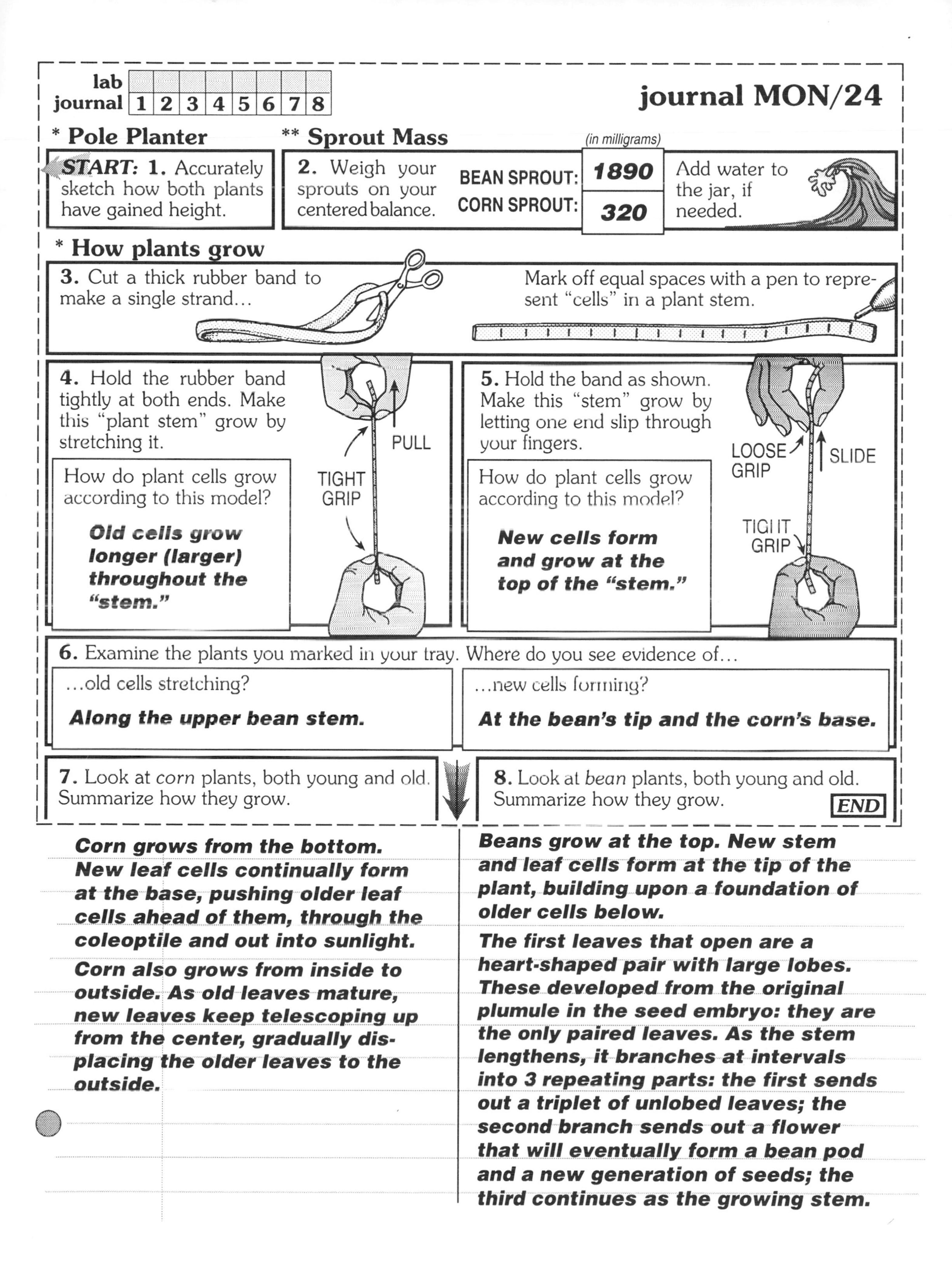

lab journal | 1 | 2 | 3 | 4 | 5 | 6 | 7 | 8

journal MON/24

* Pole Planter

START: **1.** Accurately sketch how both plants have gained height.

** Sprout Mass

2. Weigh your sprouts on your centered balance.

	(in milligrams)
BEAN SPROUT:	***1890***
CORN SPROUT:	***320***

Add water to the jar, if needed.

* How plants grow

3. Cut a thick rubber band to make a single strand… Mark off equal spaces with a pen to represent "cells" in a plant stem.

4. Hold the rubber band tightly at both ends. Make this "plant stem" grow by stretching it.

How do plant cells grow according to this model?

Old cells grow longer (larger) throughout the "stem."

5. Hold the band as shown. Make this "stem" grow by letting one end slip through your fingers.

How do plant cells grow according to this model?

New cells form and grow at the top of the "stem."

6. Examine the plants you marked in your tray. Where do you see evidence of…

…old cells stretching?

Along the upper bean stem.

…new cells forming?

At the bean's tip and the corn's base.

7. Look at *corn* plants, both young and old. Summarize how they grow.

Corn grows from the bottom. New leaf cells continually form at the base, pushing older leaf cells ahead of them, through the coleoptile and out into sunlight.

Corn also grows from inside to outside. As old leaves mature, new leaves keep telescoping up from the center, gradually displacing the older leaves to the outside.

8. Look at *bean* plants, both young and old. Summarize how they grow. **END**

Beans grow at the top. New stem and leaf cells form at the tip of the plant, building upon a foundation of older cells below.

The first leaves that open are a heart-shaped pair with large lobes. These developed from the original plumule in the seed embryo: they are the only paired leaves. As the stem lengthens, it branches at intervals into 3 repeating parts: the first sends out a triplet of unlobed leaves; the second branch sends out a flower that will eventually form a bean pod and a new generation of seeds; the third continues as the growing stem.

* **Graph your Mass Data**

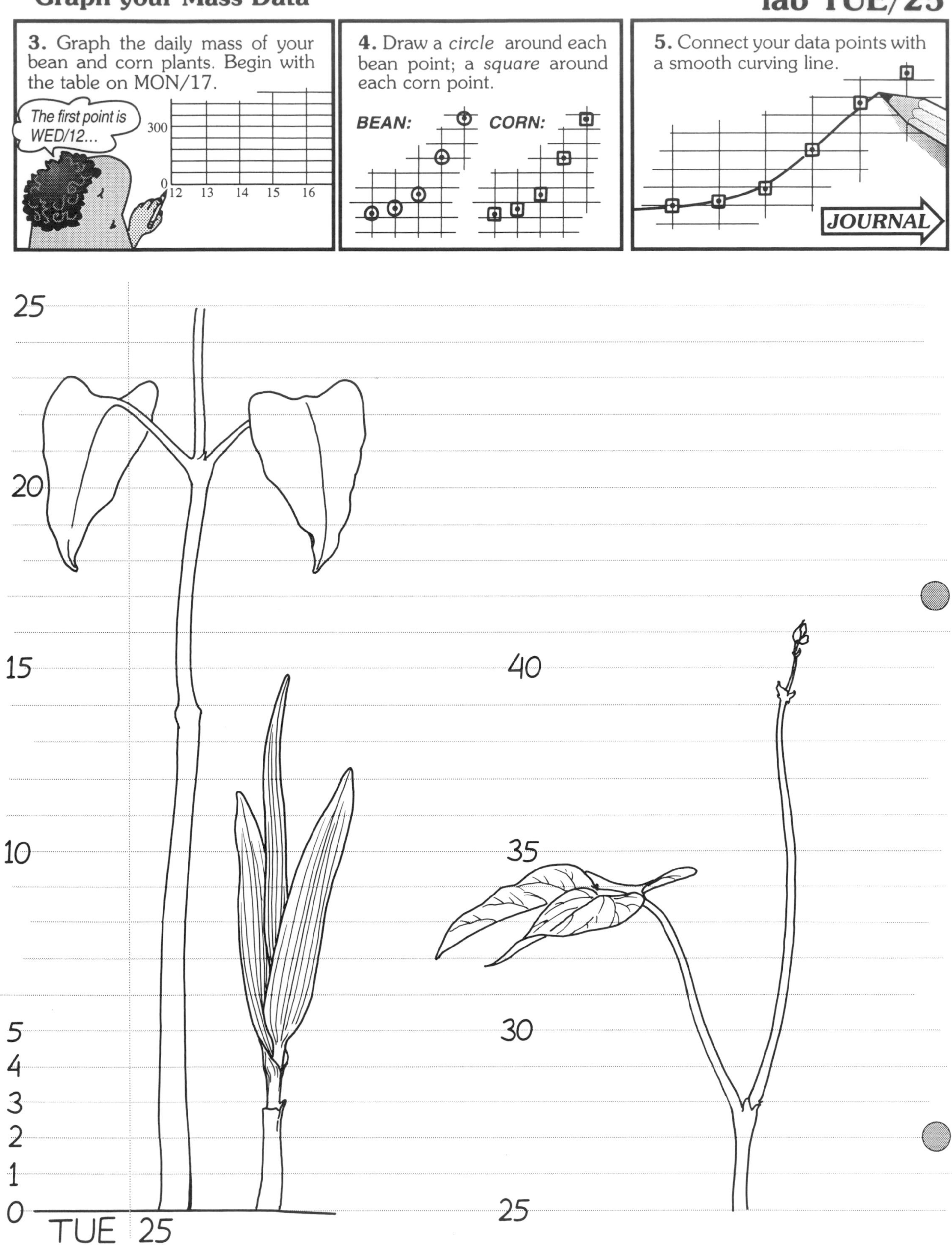

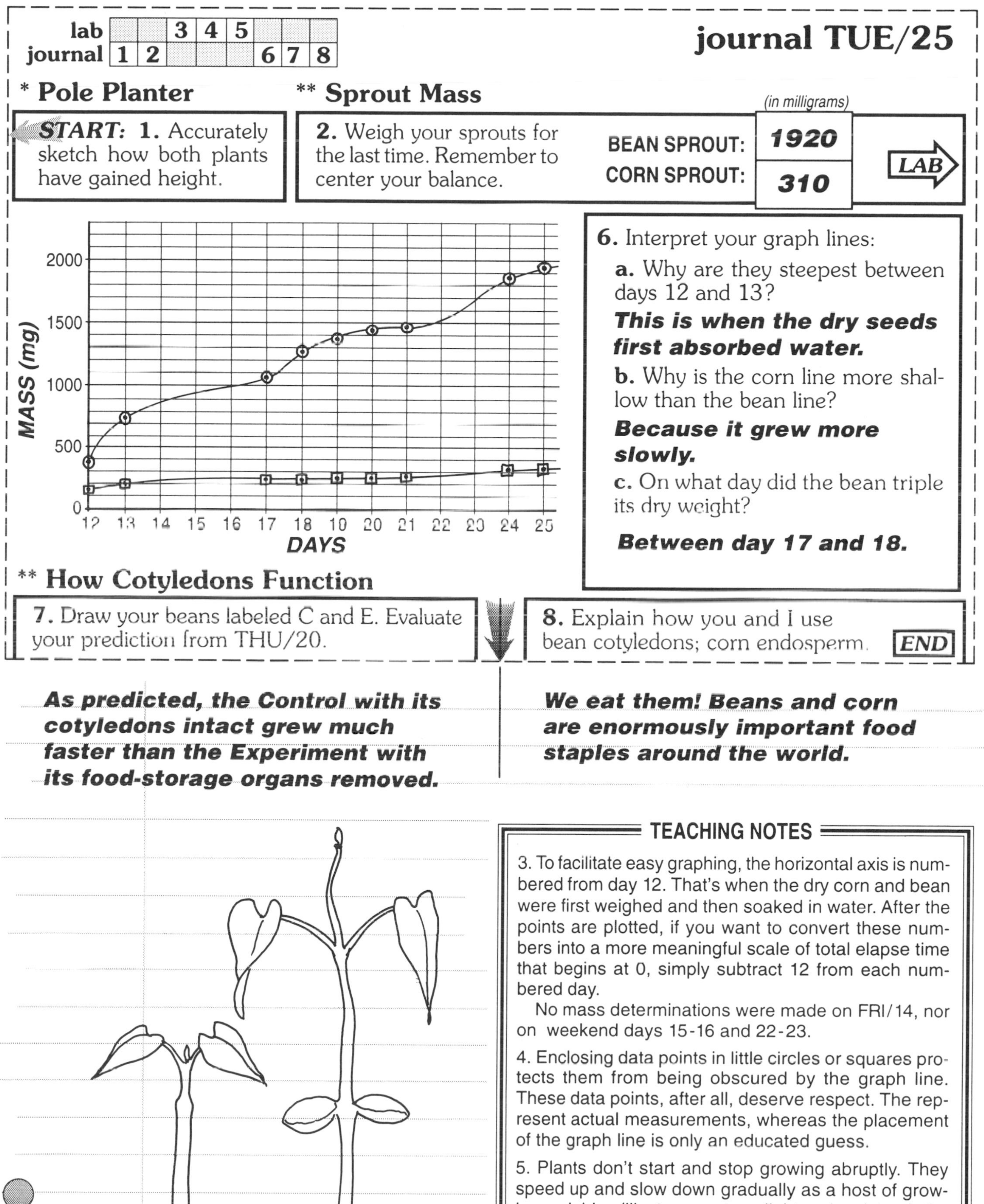

lab			3	4	5			
journal	1	2				6	7	8

journal TUE/25

* Pole Planter

START: 1. Accurately sketch how both plants have gained height.

** Sprout Mass

2. Weigh your sprouts for the last time. Remember to center your balance.

	(in milligrams)
BEAN SPROUT:	***1920***
CORN SPROUT:	***310***

LAB

6. Interpret your graph lines:

a. Why are they steepest between days 12 and 13?

This is when the dry seeds first absorbed water.

b. Why is the corn line more shallow than the bean line?

Because it grew more slowly.

c. On what day did the bean triple its dry weight?

Between day 17 and 18.

** How Cotyledons Function

7. Draw your beans labeled C and E. Evaluate your prediction from THU/20.

As predicted, the Control with its cotyledons intact grew much faster than the Experiment with its food-storage organs removed.

8. Explain how you and I use bean cotyledons; corn endosperm. END

We eat them! Beans and corn are enormously important food staples around the world.

TEACHING NOTES

3. To facilitate easy graphing, the horizontal axis is numbered from day 12. That's when the dry corn and bean were first weighed and then soaked in water. After the points are plotted, if you want to convert these numbers into a more meaningful scale of total elapse time that begins at 0, simply subtract 12 from each numbered day.

No mass determinations were made on FRI/14, nor on weekend days 15-16 and 22-23.

4. Enclosing data points in little circles or squares protects them from being obscured by the graph line. These data points, after all, deserve respect. The represent actual measurements, whereas the placement of the graph line is only an educated guess.

5. Plants don't start and stop growing abruptly. They speed up and slow down gradually as a host of growing variables (like temperature, light and moisture) dictate. Hence curves, not jagged pointy lines, best represent natural growth.

TEACHING NOTES

2. Unlike the previous graph, this one measures total elapsed time since the bean and corn were first exposed to moisture way back on FRI/0. The graph begins on FRI/7, since that is when the sprouts were first planted in the pole planter, and their heights recorded. Data points are missing for each weekend when no measurements were taken.

4-5. These experiments have been in progress for 12 days, since FRI/14. This is enough time to produce yellowing in the sunlight-deprived leaf that is covered by foil.

The leaf that is greased on the bottom, blocking intake of CO_2 through the stomata, will also yellow, but the effect is more subtle. Apparently the chlorophyll can survive a lack of gas exchange over a longer time than a lack of light, before finally breaking down.

If you leave at least one greased-leaf experiment in place until this module ends (or perhaps a week or 2 longer), this yellowing gradually becomes more apparent. (In time, all lower true leaves age and yellow, as newer leaves further up the bean plant take over the task of photosynthesis.)

Be careful about overgeneralizing from beans to all other plants. Stomata sometimes occur on the stems of plants, and on leaf tops rather than bottoms. Floating lily pads, for example, have stomatal openings that occur exclusively on the top, for a very good reason.

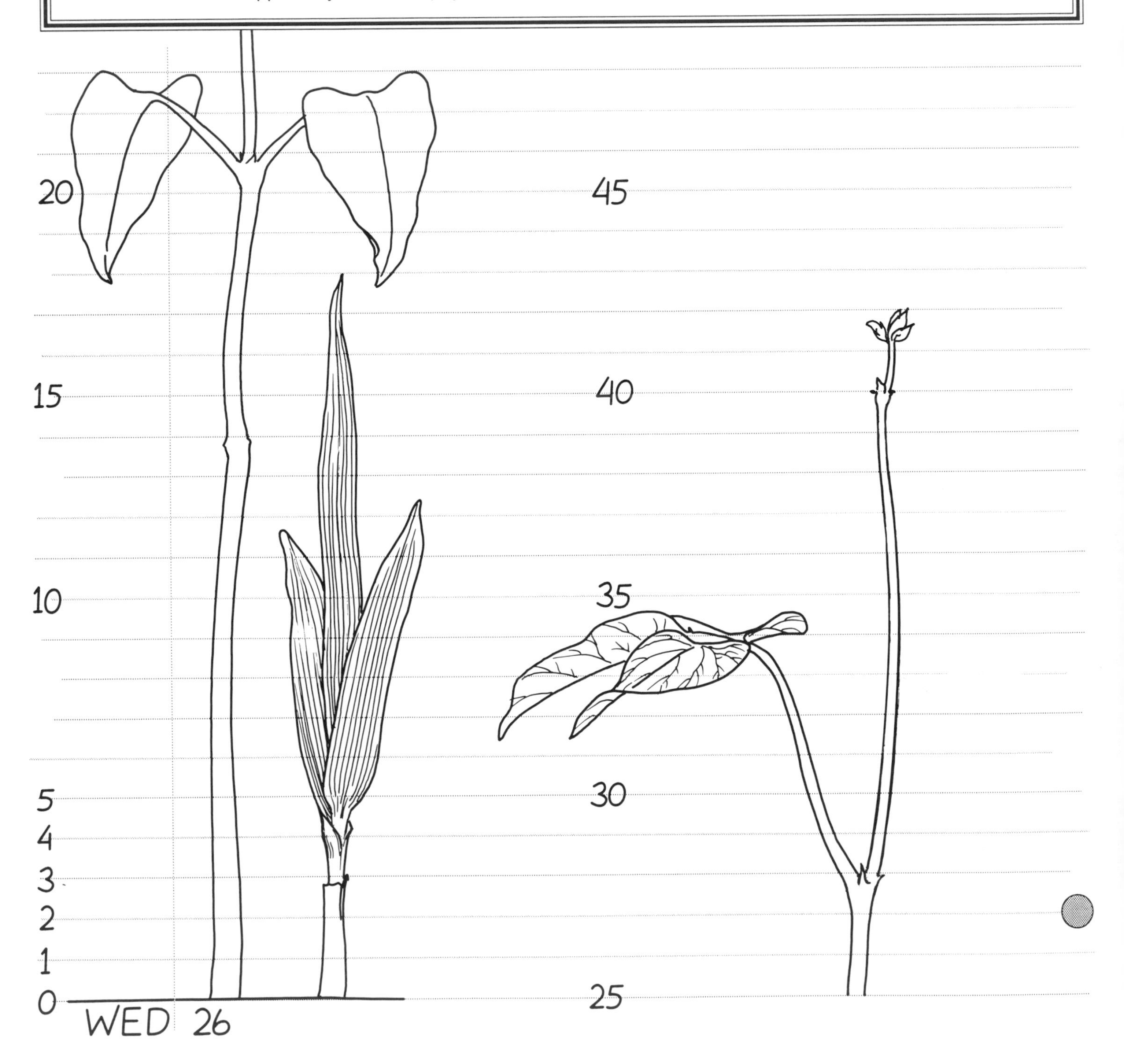

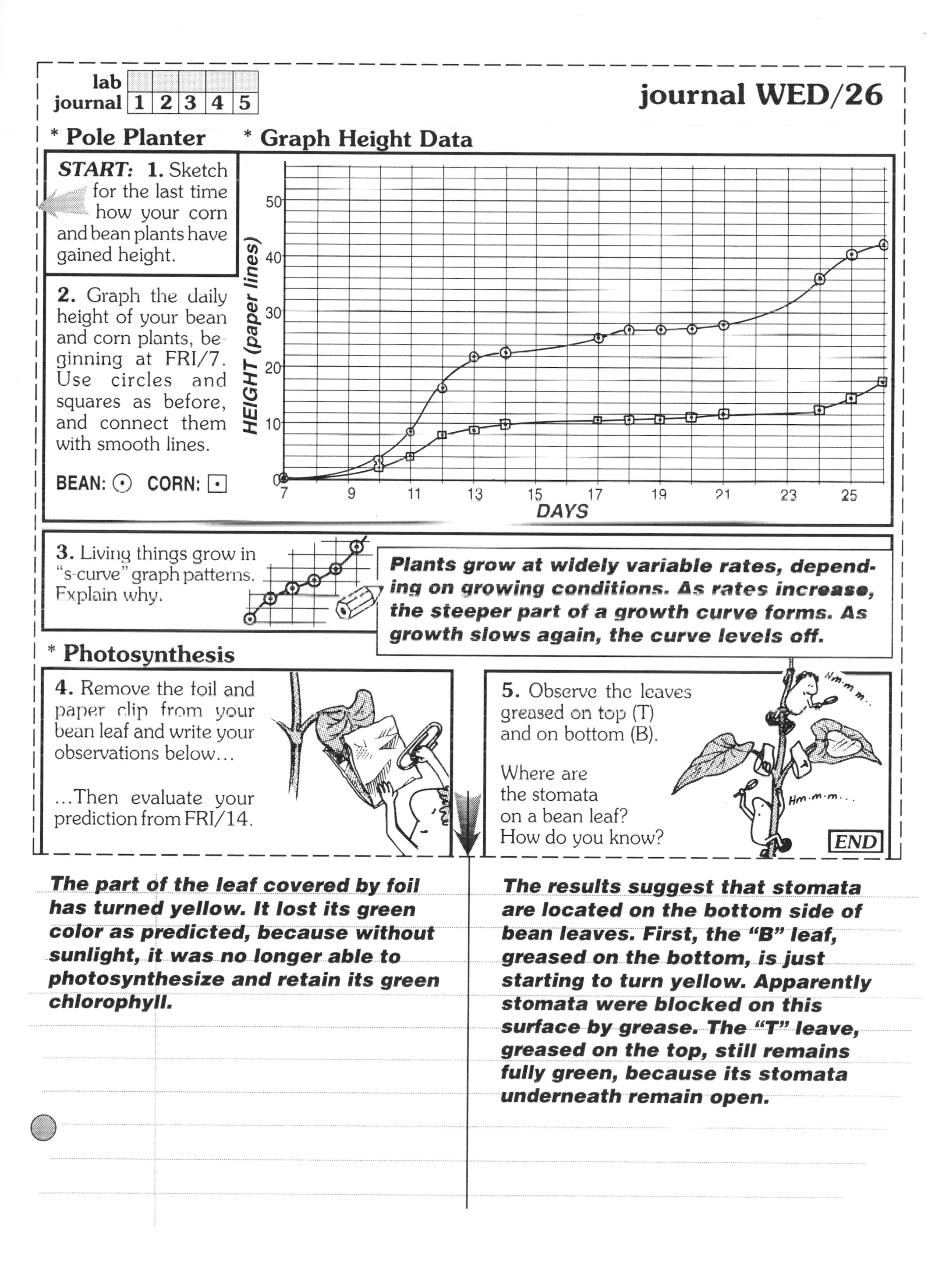

lab journal | 1 | 2 | 3 | 4 | 5

journal WED/26

* Pole Planter

START: **1.** Sketch for the last time how your corn and bean plants have gained height.

* Graph Height Data

2. Graph the daily height of your bean and corn plants, beginning at FRI/7. Use circles and squares as before, and connect them with smooth lines.

BEAN: ⊙ CORN: ⊡

3. Living things grow in "s-curve" graph patterns. Explain why.

Plants grow at widely variable rates, depending on growing conditions. As rates increase, the steeper part of a growth curve forms. As growth slows again, the curve levels off.

* Photosynthesis

4. Remove the foil and paper clip from your bean leaf and write your observations below...

...Then evaluate your prediction from FRI/14.

5. Observe the leaves greased on top (T) and on bottom (B).

Where are the stomata on a bean leaf? How do you know?

END

The part of the leaf covered by foil has turned yellow. It lost its green color as predicted, because without sunlight, it was no longer able to photosynthesize and retain its green chlorophyll.

The results suggest that stomata are located on the bottom side of bean leaves. First, the "B" leaf, greased on the bottom, is just starting to turn yellow. Apparently stomata were blocked on this surface by grease. The "T" leave, greased on the top, still remains fully green, because its stomata underneath remain open.

1-2. To explain the function of stems and flowers, your students will have to rely on their own background knowledge and experience. If they need additional input, you might write brief definitions on your blackboard, discuss stem and flower functions in advance, or send students to the library.

SOAK LENTILS AND WHEAT BERRIES OVERNIGHT

Students will begin tomorrow's take-home test by finding the average mass of lentils and wheat berries, both dry and presoaked. They should do this first thing, since FRI/28 may be the last day that balances are still available to all students. Soak perhaps 150 wheat berries and 150 lentils right now (6 big pinches each), in separate baby food jars filled with water. Set aside until tomorrow.

JOURNAL EVALUATIONS

Today's schedule is light enough to allow time to grade plant journals. If you have initialed your approval of student work on a daily basis, simply count your signatures and assign journal grades based on total number.

Otherwise, collect all journals at the end of the period to evaluate and return by tomorrow, FRI/28. Your students will use these journals as references when completing their take-home tests. In addition, they will be eager to show family members their completed work.

We recommend that journal grades comprise perhaps a third of each student's overall science grade for Corn and Beans. The take-home test might form another third of the grade, with daily work habits and attitude completing the final third.

Students who finish early might get a head start on tomorrow's take-home test by finding the mass of 10 dry lentils and 10 dry wheat berries. With this information and a basic understanding of decimal division, they should be able to complete these 4 equations on scratch paper.

10 dry lentils = _____mg
1 dry lentil = _____mg
10 dry wheat berries = _____mg
1 dry wheat berry = _____mg

Another way to use extra time is to suggest that students decorate their journal covers in crayon or colored pencil.

WHAT TO DO WITH ALL THOSE CORN AND BEANS

Brainstorm what to do with the many corn and bean plants still growing in your classroom. Do this now, if time remains, or tomorrow after students have completed their mass determinations on the take-home test. You may wish to include some of the these options in your discussion:

❦ **Enough already.**

Compost all the corn and beans outside in a flower bed or school garden. Discuss how their elements will recycle into new forms. (This option is quick and easy. Baby food jars are free to use for the take-home test, and you reserve jars, cans and other useful items for next year when you teach this unit again.)

❦ **Take everything home.**

Your room may already be overstocked with jars and cans, and certain students may now be pleading to take their plants home. To exercise this option, ask students to peacefully decide among themselves who keeps what. Once this is accomplished, distribute large paper grocery bags to organize and transport materials. Direct students to prepare these bags as follows:

(a) Cut off the top third of a folded bag, equal to the width of its bottom. Lay this cutoff in the bottom of the bag, which is now about $2/3$ of its original height, with a lined bottom.

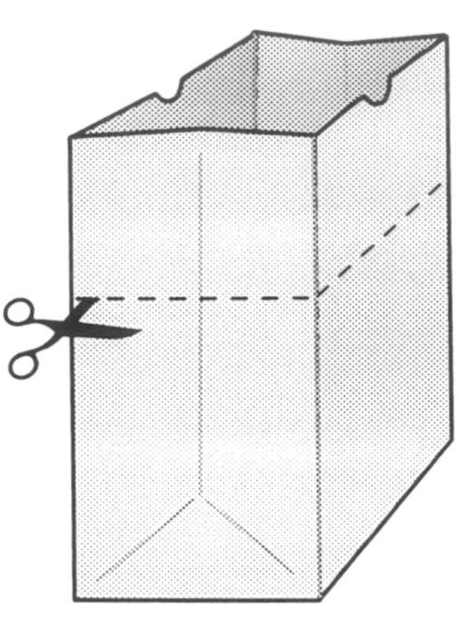

(b) Fold half the remaining height of the side "walls" to the inside. The bag is now $1/3$ of its original size, with lined sides and bottom.

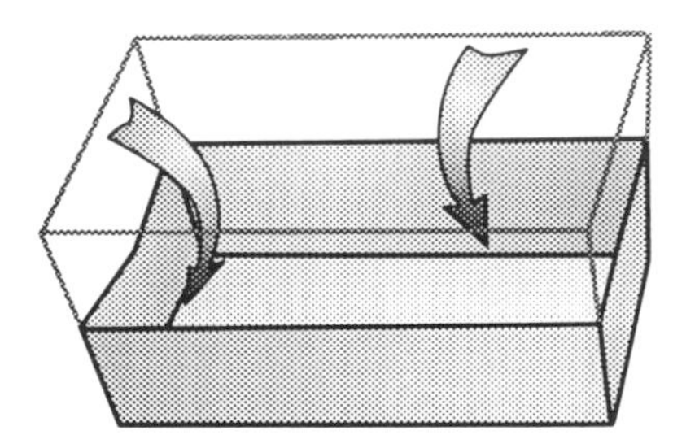

You now have a lightweight, reinforced "box."

(c) Empty the pint or quart jar, and wring excess water from the wicking towels. Arrange the equipment inside the bag. Paper-clip the pole planter to the side of the bag so it won't tip over.

❦ **Continue group experimentation.**

Reserve some (or all) of your plants for further study and observation. Transplant equal numbers of corn and beans into various mediums and compare growth responses.

(a) Maintain a control group growing <u>without nutrients</u> in moist vermiculite. If mold is present, clean and disinfect the plant tray, then transplant into fresh vermiculite.

(b) Transplant a second group into an equal amount of <u>potting soil</u>. Clean and disinfect the plant tray, as necessary.

(c) Experiment with various concentrations of <u>plant nutrients</u>: nitrogen, potassium and phosphorous.

(d) Transplant into a <u>large volume of soil</u>, either a large inside planter or, weather permitting, an outside garden plot.

Given favorable conditions, some beans will complete a full life cycle — from seed to sprout, to mature plant, to self-pollinating flowers, to pod, and finally back to seed, which you can save and plant next year.

The corn may not reach maturity unless planted outside. Since corn is wind-pollinated, several rows of plants are needed to achieve well-developed ears.

lab journal 1 2

journal THU/27

* Examine Mature Plants

START: 1. This is how each plant looks fully grown. Write the correct label next to each part.

adventitious root
primary root
node
leaf
stalk
flower
fruit

CORN:

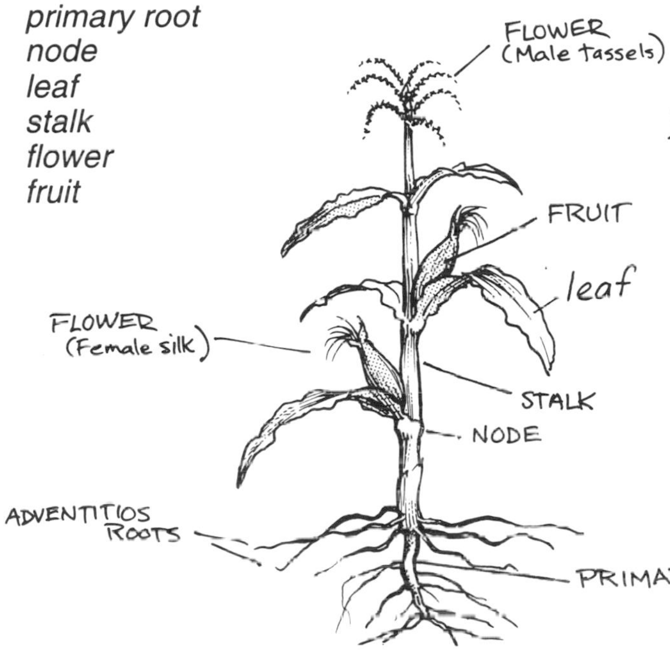

FLOWER (Male tassels)
FRUIT
leaf
FLOWER (Female silk)
STALK
NODE
ADVENTITIOS ROOTS
PRIMARY ROOT

BEAN:

primary root
secondary root
hypocotyl
first leaves
leaf triplet
node
stem
flower
pod

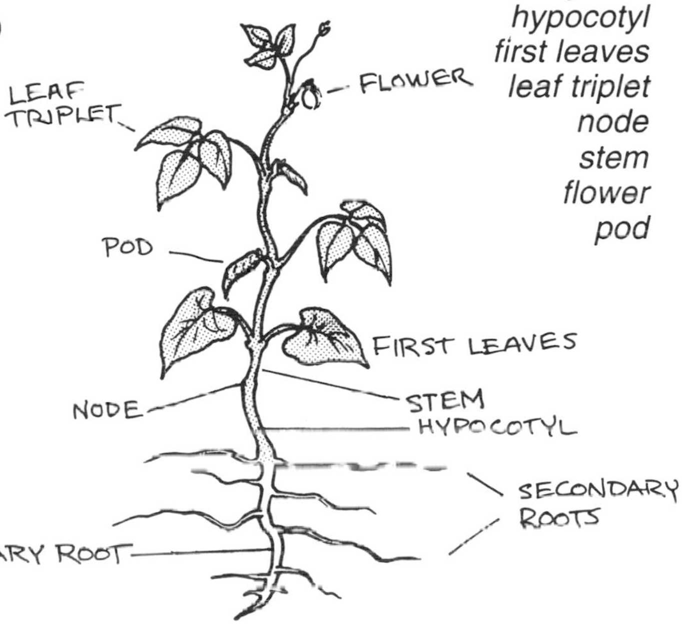

FLOWER
LEAF TRIPLET
POD
FIRST LEAVES
NODE
STEM
HYPOCOTYL
SECONDARY ROOTS

2. Divide your 2 columns below into 4 boxes labeled ROOTS, STEMS, LEAVES and FLOWERS. Explain how plants use each part.

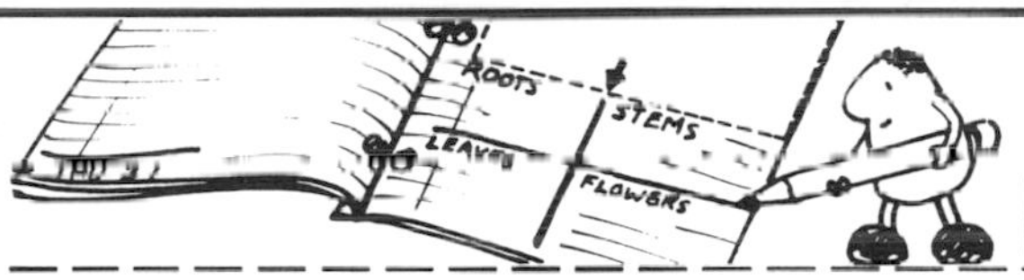

ROOTS STEMS LEAVES FLOWERS

END

ROOTS:
These absorb water and minerals into the plant, anchor it firmly in the ground, and store food reserves derived from photosynthesis.

STEMS:
These provide a channel to transport water and minerals up from the roots, and food reserves down from the leaves.

LEAVES:
These combine water and carbon dioxide in the presence of sunlight to photosynthesize food for the rest of the plant.

FLOWERS:
These reproduce the plant by forming the next generation of seeds.

BEGIN THE TAKE-HOME TEST

Follow these simple steps to prepare your class for a home study of lentils and wheat:

1. Display dry lentils and wheat berries in two labeled containers, as well as your presoaked seeds from yesterday.

2. Distribute a take-home test to each student. Ask lab groups to first complete the mass equations in part (c). Emphasize that these seeds need only to be weighed in groups of 10. The mass of single seeds is found by simply moving the decimal point.

If you don't have enough class time for students to make their own mass determinations, have them copy this data into box (c) of their test papers:

10 dry lentils = 480 mg
1 dry lentil = 48 mg
10 dry wheat berries = 310 mg
1 dry wheat berry = 31 mg
10 soaked lentils = 1,020 mg
1 soaked lentil = 102 mg
10 soaked wheat berries = 400 mg
1 soaked wheat berry = 40 mg

3. Put 1 big pinch each of dry lentils and wheat berries in a baby food jar and cover with a lid. (There are 20 to 30 seeds in a big pinch.) Distribute one jar to each student.

4. Review the entire take-home test with your whole class. Discuss your performance expectations, and answer any questions students may have. Establish a due date, 2 weeks from today, and hold to it.

CLEAN UP

If you have not already done so, brainstorm with your class how you might use the corn and beans still growing in your classroom. (See teaching notes THU/27.) After you reach a class consensus, clean up and reorganize your classroom accordingly.

EVALUATE THE TAKE-HOME TEST

Expect a variety of student output. Award grades based on total effort and quality of work. Following are some fruitful areas of investigation that parallel the journal work students have already completed:

• A periodic record, both written and visual, of how wheat and lentils develop over the 2 week period.
• Statements of comparison or contrast about dry seeds, moist seeds, germination, growth of sprouts, development of leaves and roots.
• A quantitative analysis of water absorption.
• The classification of wheat as a monocot and lentils as a dicot. Students should refer to similarities between wheat and corn; between lentils and beans.
• Experiments involving any of these topics: photosynthesis, the location of stomata on leaves, cotyledons, cell growth, graphing height as a function of time.

TYPICAL GROWTH of LENTILS and WHEAT, actual size:

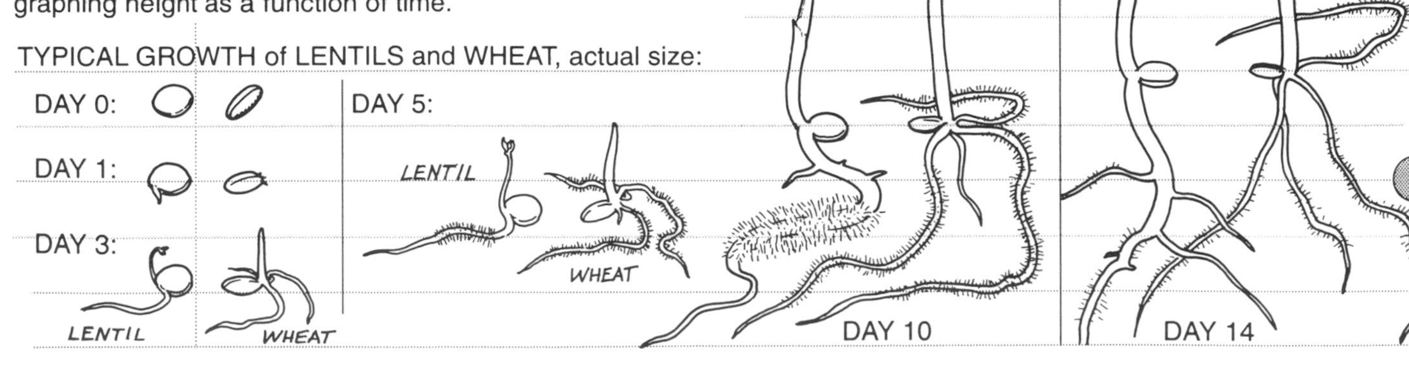

Dear Family Member,

Over the next 2 weeks your son or daughter will study how seeds grow, applying concepts and skills already presented and practiced in class. Beyond paper towels and small jars, no special equipment is required.

Please participate. Great leaps forward in learning could result! Keep in mind that this is a test. Assist without directing; encourage independent thought and action.

Dear Student,

You are given wheat berries and lentils in a sealed baby food jar.

Your task is to find out all you can about each kind of seed by doing experiments. Please report your findings in an organized way, using complete sentences and detailed drawings.

You have 2 weeks. Do your best job. These instructions will help get you started.

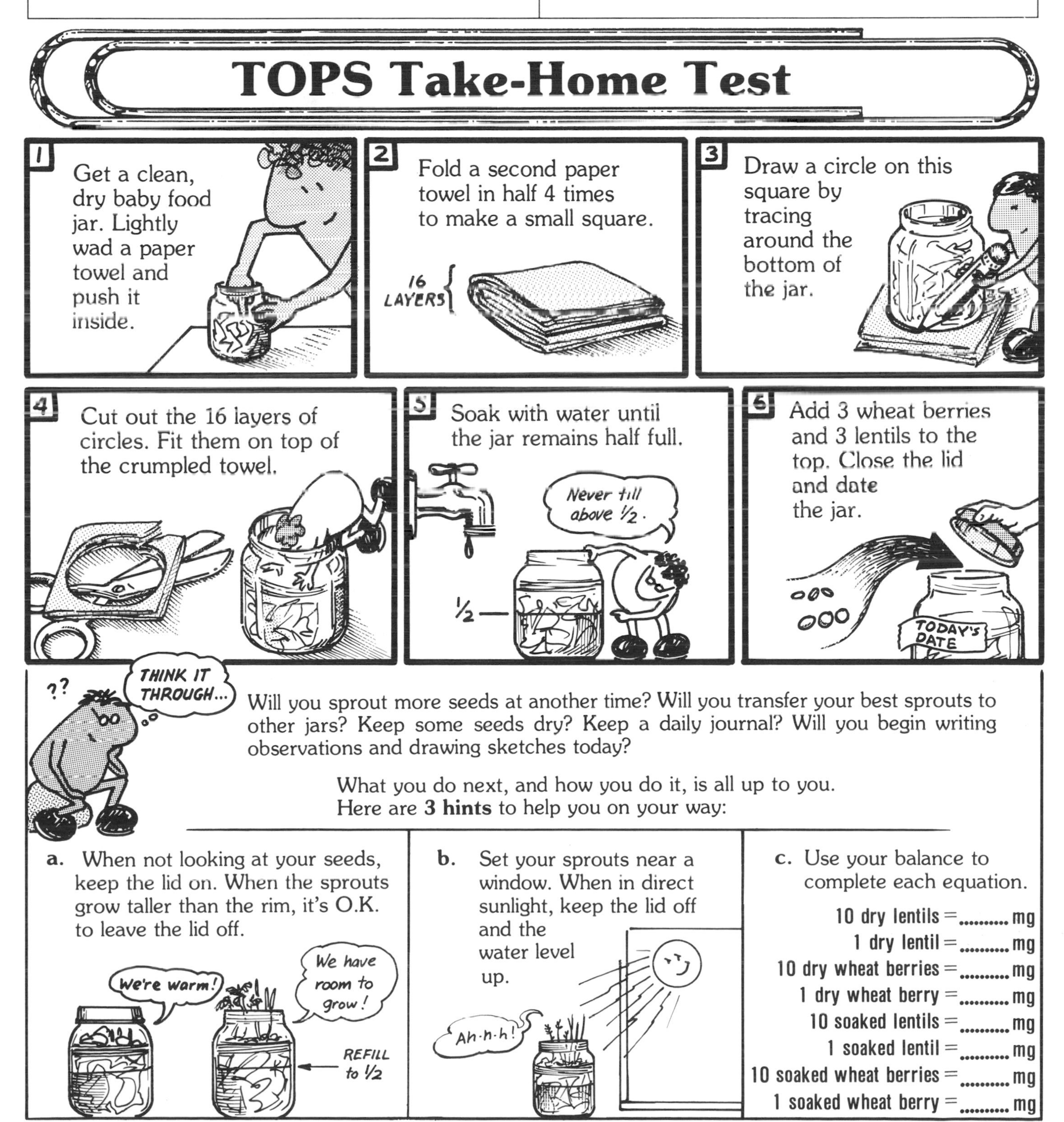

LAB INSTRUCTIONS

5 weeks
11 pages

CALENDAR

Begin this page on **MON/-4**, the first day of the first week.
You'll finish these lab instructions 5 weeks later, on **TUE/25**.

First Week:	**MON/-4**	TUE/-3	WED/-2	THU/-1	FRI/0
Second Week:	MON/3	(journal)	WED/5	THU/6	FRI/7
Third Week:	MON/10	(journal)	WED/12	(journal)	FRI/14
Fourth Week:	MON/17	TUE/18	(journal)	THU/20	FRI/21
Fifth Week:	(journal)	**TUE/25**	(journal)	(journal)	

* Make your Plant Journal

lab MON/-4

START: **1.** Get 12 journal cutouts from your teacher.

2. Cut around the dashed lines into 23 separate parts:

3. Paper clip the four paper masses together, and set them aside.

4. Tape each weekday to a new sheet of notebook paper, even with the upper right corner.

5. Arrange all 19 journal days in order, beginning with MON/3. Add a cover sheet.

6. Tap the edges even. Staple very near each hole like this:

7. Crease each page open along the spine, and extend the arrows to the bottom of each page.

8. Write "PLANT JOURNAL" on the cover of your book. Put your name in the lower right corner.

9. Tuck your paper masses inside.

END

** Build a Balance

10. Trace around this large dashed box with 2 pieces of foil underneath….
Cut out both foil rectangles.
11. Fold 1 piece in half the long way…
Wrap it around a battery as shown…
FOLDED EDGE
BUMP END
HALF OFF BATTERY
12. Fold over the foil ends. Push the bottom flat on the table to make a "bucket"….
FOLD OVER
PUSH FLAT
…Remove the battery.
13. Fold the other piece into quarters.
FOLD
FOLD
14. Hang one of these foil pieces from each paper clip at the ends of your straw.
15. Fold a small piece of masking tape almost in half….
STICKY END
Stick this "rider" on the high side of the straw so it balances level.
16. Get the paper masses you tucked inside your journal. Cut out each one, staying exactly on the outside solid lines.
Be really careful!
17. Fold the 3 largest masses in half. Leave the smaller ones flat.
FOLD TO SHORTEN
LEAVE OTHERS FLAT
18. If your are working with a lab partner, you will have an extra set of masses. Save these in a safe place to replace ones that are damaged or lost.
Use these…
…save these.
19. Find the mass of a bean, corn seed, and paper clip….
A. Shift rider so straw is level when balance is empty.
B. Put bean, corn seed or paper clip here.
C. Slide paper masses between paper clip and foil.
D. Add masses until straw balances level.
Compare your answers with each other and the teacher.
20. Paper clip your masses. Store them inside the can.
END

** Make a Pole Planter
lab WED/-2
START: 1. Pull up the middle of a paper clip...
...until you make it straight.
whew!
Wrap the big end in masking tape.
2. Roll up masking tape so it is sticky on the outside...
ABOUT THIS BIG
Stick it to the outside of a baby food jar.
3. Fix the wrapped end of your paper clip to the sticky patch, then tape over the top.
Write your name(s) on the tape.
Name(s)
4. Pull the "arm" of the paper clip out just enough to fit a straw snugly over the top.
5. Trim 2 sheets of notebook paper along the top and bottom lines.
6. Fold these in half 4 times, to make narrow strips as wide as your little finger.
7. Clear-tape each strip in the middle and at both ends.
8. Cut off half the straw you stuck onto your jar.
9. Use this piece to join both tubes of notebook paper together.
HALF STRAW
10. Number up this long strip (starting with "0") so each line runs through its number.
GAP
START FROM "0"
Number across the gap, too.
11. Stick this numbered strip over the half straw on your jar.
This makes a BEAN POLE!
12. Cut masking tape as long as your finger. Fold it lengthwise, sticky sides together.
FOLD
Make cuts from opposite sides about here:
13. Tape it to the back of your pole at line 20:
TAPE
Use it later to tie up your growing bean.
CONTINUE

** Number Four Bottle Caps
14. Trace around a bottle cap on scratch paper to make four circles. Number them boldly 1, 2, 3 and 4.
15. Cut out the 4 numbered circles, and press them into 4 bottle caps.
16. Get a deli-tub. Label it with your name(s) on a piece of masking tape.
ELI
** Make a Storage Mat
17. Store these bottle caps, plus a folded paper towel, inside your labeled tub.
18. Divide and label a sheet of paper like this:
Name(s)
MILK CARTON TRAY
DELI TUB
BALANCE
POLE PLANTER
19. Store your lab equipment on this space-saving mat.
Name(s)
NAMES
NAMES
END
* Make a Bean Map
lab THU/-1
START: 1. Cover this square with scratch paper. Carefully trace the square and the beans.
SCRATCH PAPER
2. Cut out your square. Fold it in the middle to make a stand-up "bean."
My BEAN MAP!
3. Choose an average-looking bean that has medium size and normal color.
Pretty ordinary...
4. Carefully draw all the tiny seed coat patterns on both sides of your special bean.
SAVE YOUR BEAN!
5. Mix your special bean with 10 others. Use your map to find it again...
Can you find it in a group of 50?
6. Trade maps and beans. Can a friend find your special bean in a group of 50?
7. Tape your bean map inside your journal cover.
END

** Make a Milk Carton Seed Tray

* Begin your Journal

lab MON/3

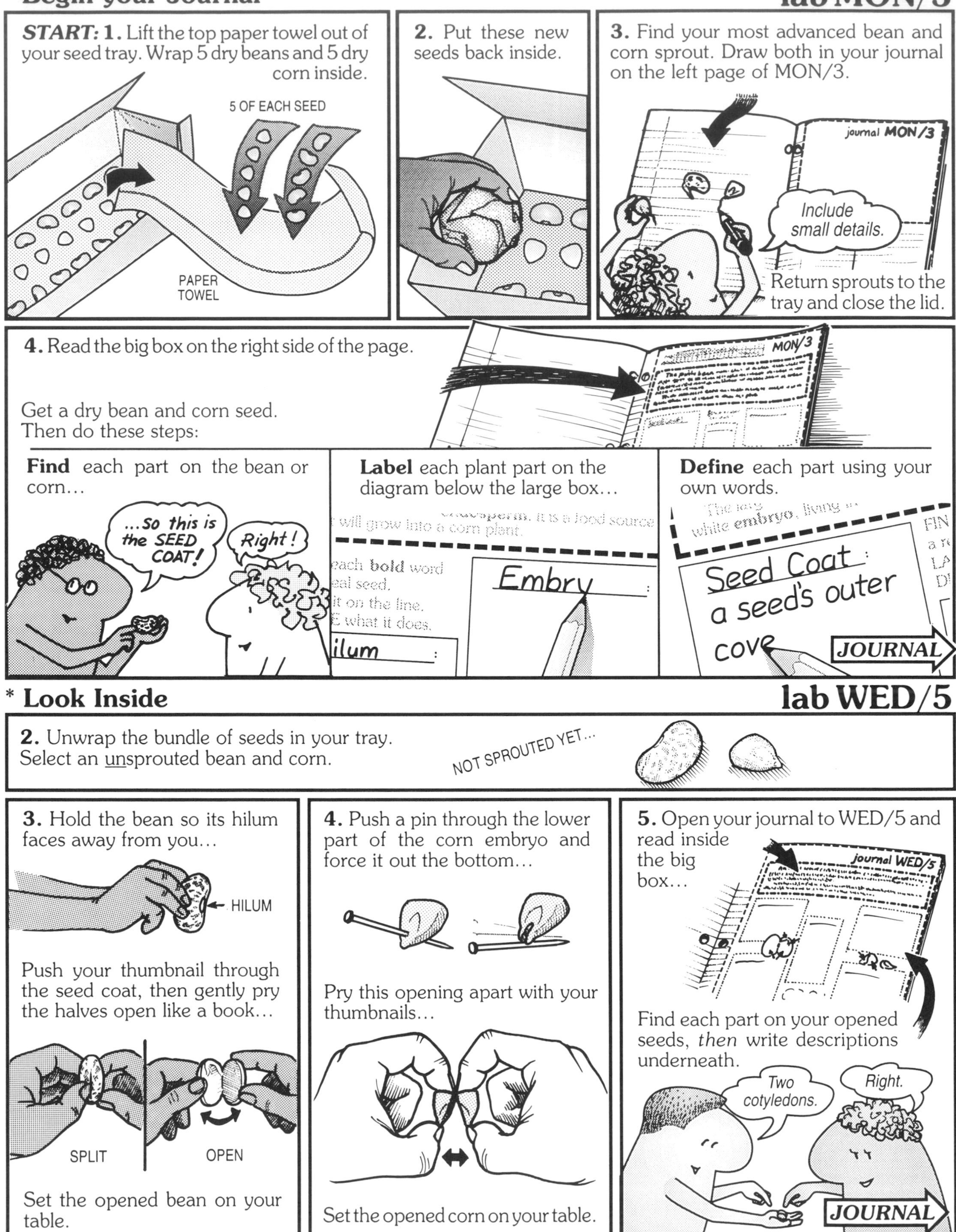

** Plant your Pole Planter
lab THU/6
2. Cut a piece of paper towel about as big as this rectangle...
Roll it lengthwise into a tube, then moisten with water.
Press it inside your pole planter like this. Hold it with a rubber band.
RUBBER BAND
3. Fill with vermiculite.
GREEN THUMB
VERMICULITE
4. Slowly add water, then drain completely.
5. Poke 2 deep pencil holes into the jar.
6. Choose a bean sprout and a corn sprout from your seed tray with roots that fit the holes...
Rest them in the holes so the tops just stick out.
Push soil gently against the roots.
* Look at Growing Sprouts
7. Open your journal to THU/6 and read inside the big box.
journal THU/6
Find each part on real sprouts before you write your descriptions.
JOURNAL
* Prepare your Journal for Daily Drawings
lab FRI/7
START: 1. Open your journal to the left side of FRI/7.
FRI/7
2. Adjust your pole so "0" is even with the soil surface...
EVEN!
Draw a heavy line to represent this surface, and write "FRI/7" beneath. Number up the edge of the page from "0" to "5" so each line passes through its number.
SURFACE
FRI / 7
BOTTOM LEFT CORNER of JOURNAL
3. Draw how far each sprout pokes above line "0" (the ground).
The bean is a little above the "0".
FRI/7
4. Draw similar "0" lines in the lower left corner of all remaining journal pages. Write the correct days, and numbers up the sides, as before.
Number every 5th line past 5.
Do this to all pages through WED/26.
TUE /11
MON /10
FRI/7
JOURNAL

** Plant your Seed Tray
lab MON/10
4. Cut the lid off your tray and divide it into 4 long, even strips: save these. Lift out the damp towel with all the sprouts growing on it, and set it on your table.
CUT LID INTO STRIPS
BE GENTLE!
5. Discard the remaining 2 towels. Thoroughly clean both tray and jar, then reassemble with 3 new towels. Moisten to lay flat, but don't yet fill the jar with water.
Fresh and clean!
NO WATER
6. Fill the tray 1/4 full with vermiculite. Add enough water so the particles *just begin* to float, forming a soft, wet ooze.
SOFT, THICK SLURRY
1/4 FULL
JAR CATCHES OVERFLOW
7. Gently pull your 10 best plants from the towel and poke them into the slurry as shown. Plant 9 new seeds as well.
4 CORN
6 BEAN
6 DRY BEANS + 3 DRY CORN
Set the extras aside.
FREE SPROUTS
1/4 TRAY
1/2 TRAY
1/4 TRAY
8. Drain excess water from the tray (if necessary) by pushing the towels to one side. Then fill the jar 3/4 full.
FILL 3/4 FULL.
9. Paper clip the strips cut from the lid onto your seed tray to make 2 arches. Connect these arches with masking tape to support your growing plants.
2 STRIPS MAKE EACH ARCH
Like a covered wagon...
PLANT SUPPORTS
END
** Weigh Numbered Seeds
lab WED/12
2. Push 4 dry beans and 2 dry corn deep into your tray, and cover them over.
3. Take the numbered bottle caps out of your deli-tub. Moisten the towel so it lies flat in the tub.
4. Set the bottle caps on the moist towel, and fill each one with water.
CONTINUE

5. Get a dry bean. Find its mass on your balance.
Center the balance first so the straw is level.
RIDER
Record its mass in the #1 column of your journal, WED/12…
#1
DRY BEAN: 340
DRY CORN: 140
Soak this bean in the #1 bottle cap.
Repeat these steps with a dry corn. Pair it with your #1 bean.
6. Continue until you fill the journal table and bottle caps with four pairs of seeds.
JOURNAL
** Add Water
** Cover Some Bean Leaves
lab FRI/14
2. Check your plants. Water them if necessary. Is your pole planter wick moist?
Is the water jar under your plant tray at least half full?
3. Trace this square onto a piece of foil.
Cut it out…
…then fold it into quarters.
4. Find a large bean leaf in your plant tray and slip the foil over it. Hold it on with a paper clip.
5. Divide a short piece of masking tape into 2 narrow strips. Label one "T" for top, the other "B" for bottom.
CUT LENGTHWISE.
T
B
6. Tag another pair of large bean leaves with each letter.
7. Smear some grease across only the bottom of leaf "B", and only the top of leaf "T."
ONE SIDE ONLY!
JOURNAL
** Tie a Bean and Corn Sprout
lab MON/17
2. Open your deli-tub and remove your most advanced bean sprout. Gently tie a thread on its hypocotyl.
DOUBLE KNOT
3. Tie another thread around your most advanced corn sprout, just behind its coleoptile.
Record both seed numbers on scratch paper.
Trim one end of each thread short: leave one end as long as your hand.
4. Write the bottle cap number of each sprout in your journal table…
(all data in milligrams)
Wed/12
Thu/13
BEAN: #4
CORN: #
m-m-m-m-m— Snacks?!
Discard your remaining sprouts.
JOURNAL
free sprouts

* Compare Leaves
lab TUE/18
9. Remove a leaf from each uprooted plant to make pencil rubbings. Label each shape.
LEAVES UNDER PAPER
RUB LIGHTLY WITH SIDE OF LEAD
CORN
10. Draw the veins in darker, in careful detail. Store these drawings in your journal.
PLANT JOURNAL
11. Assignment: Bring leaves of each type to make more rubbings tomorrow.
PARALLEL VEINS:
BRANCHING VEINS:
Please remember!
END
** Study Cotyledons
lab THU/20
3. Divide a short piece of masking tape lengthwise. Label one "E" for experiment, the other "C" for control.
CUT IN HALF.
E
C
4. Find 2 young bean sprouts of equal size, with newly opened cotyledons. Tag each plant around its base.
E
C
5. Gently break off the cotyledons from plant "E" only. Set them in the box labeled "YOUNG cotyledons" in your journal.
sorry!
!
E
C
JOURNAL
** Study Plant Growth
lab FRI/21
4. Cut some notebook paper exactly 4 lines long, and as wide as the margin.
1
2
3
4
5. Tape a pin over the middle line so its "head and neck" clear the edge.
6. Fold in along the outside lines to make the paper exactly 2 lines wide.
BLUE LINE
7. Put one drop of blue food coloring into a lid for dipping.
BLUE
JOURNAL
* Graph your Mass Data
lab TUE/25
3. Graph the daily mass of your bean and corn plants. Begin with the table on MON/17.
The first point is WED/12...
300
0
12
13
14
15
16
4. Draw a circle around each bean point; a square around each corn point.
BEAN:
CORN:
5. Connect your data points with a smooth curving line.
JOURNAL

lab	1	2	3	4		
journal				4	5	6

journal MON/3

4. The **BEAN** is covered by a speckled **seed coat**. It protects the plant embryo, wrapped inside, from insects and water loss. The scar where the bean was once attached to its pod is called the **hilum**. Next to the hilum is a tiny hole called the **micropyle**. The embryo absorbs water most rapidly through this opening.

The largest part of the **CORN** is a yellow fruit called the **endosperm**. It is a food source for the white **embryo**, living inside, that will grow into a corn plant.

Seed coat :

FIND each **bold** word on a real seed.
LABEL it on the line.
DEFINE what it does.

DRAW LINES DOWN AT ALL ARROWS

5. List below 4 ways your dry bean and corn seeds are **different**.

6. List below 4 ways your dry bean and corn seeds are **similar**.

END

lab				4	5	6	7	8	9
journal	1	2	3						

journal MON/10

* Pole Planter

***START:* 1.** Get your pole planter. On the page to the left, carefully draw how high both seedlings now reach up the pole.

2. Next to your drawing, explain why the bean's **hypocotyl** forms a hook; why the corn's leaves hide inside a **coleoptile**.

* Draw your Sprouts

3. Accurately draw your most advanced bean and corn sprout in the space below. Label all parts.

CORN:

BEAN:

LAB

lab journal | 1 | 2 | 3 | 4 | 5 |

journal TUE/4

* **Draw your Sprouts**

***START:* 1.** Draw your fastest growing bean and corn sprout on the left page.

2. Unwrap your 10 seeds soaking in the wet towel since MON/3. Set dry corn and bean seeds beside them on your table top.

DRY

SOAKED

3. Describe these before/after changes, using *complete* sentences:

DRY SEEDS ...**before** adding water	**SOAKED** SEEDS ...**after** 24 hours
color of beans:	color of beans:
color of corn:	color of corn:
texture of beans:	texture of beans:
texture of corn:	texture of corn:
mass of beans:	mass of beans:
mass of corn:	mass of corn:
size of beans:	size of beans:
size of corn:	size of corn:

mm: 0, 10, 20, 25

4. A good picture is worth a thousand words. Draw the changes you have described above in each box.

DRY BEAN → SOAKED BEAN

DRY CORN → SOAKED CORN

5. Return your 10 moist seeds to the seed tray.

END

lab		2	3	4	5		
journal	1				5	6	7

journal WED/5

*** Draw your Sprouts**

START: **1.** Draw your fastest-growing bean and corn sprout on the left page. *LAB*

5. The **BEAN** easily breaks apart into **two cotyledons**. These store starch and protein for the developing embryo, enabling its **plumule** to grow into the first 2 true leaves and its **radicle** to develop into the roots and lower stem.

The **CORN** does not divide easily because it has only **one cotyledon**. This structure, surrounding the embryo, absorbs starch and protein from the yellow **endosperm**.

FIND each **bold** word on a real seed.
LABEL it on the line.
DEFINE what it does.

____________________:

____________________:

____________________:

____________________:

____________________:

"MONO" means "ONE"...

"DI" means "TWO."

6. Which seed above is called a "monocot" (monocotyledon)? Explain below.

7. Which seed is called a "dicot" (dicotyledon)? Explain below. *END*

PAPER MASS: **500 mg**

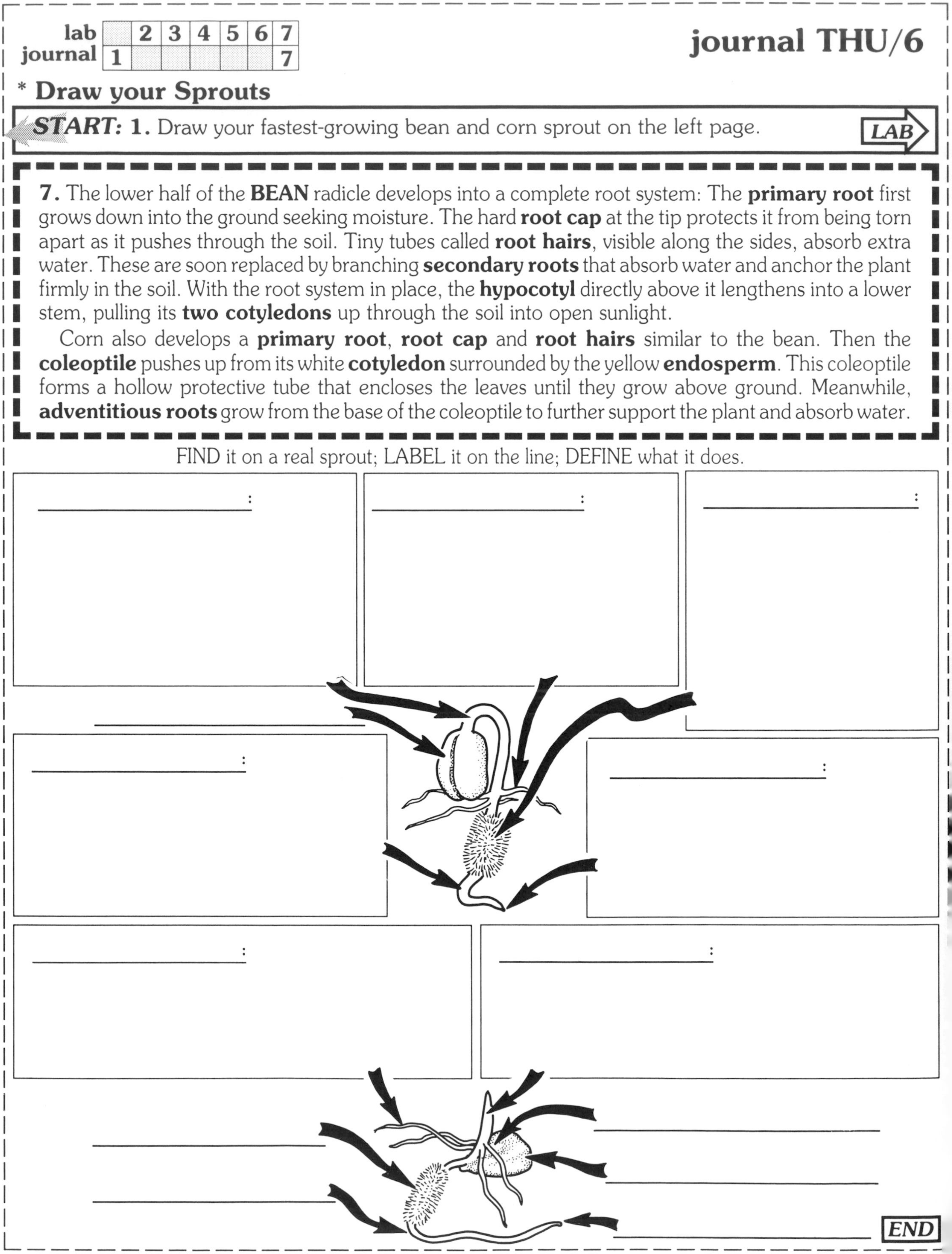

lab		2	3	4	5	6	7
journal	1						7

journal THU/6

* **Draw your Sprouts**

***START:* 1.** Draw your fastest-growing bean and corn sprout on the left page. *LAB*

7. The lower half of the **BEAN** radicle develops into a complete root system: The **primary root** first grows down into the ground seeking moisture. The hard **root cap** at the tip protects it from being torn apart as it pushes through the soil. Tiny tubes called **root hairs**, visible along the sides, absorb extra water. These are soon replaced by branching **secondary roots** that absorb water and anchor the plant firmly in the soil. With the root system in place, the **hypocotyl** directly above it lengthens into a lower stem, pulling its **two cotyledons** up through the soil into open sunlight.

Corn also develops a **primary root**, **root cap** and **root hairs** similar to the bean. Then the **coleoptile** pushes up from its white **cotyledon** surrounded by the yellow **endosperm**. This coleoptile forms a hollow protective tube that encloses the leaves until they grow above ground. Meanwhile, **adventitious roots** grow from the base of the coleoptile to further support the plant and absorb water.

FIND it on a real sprout; LABEL it on the line; DEFINE what it does.

______________:

______________:

______________:

______________:

______________:

______________:

______________:

END

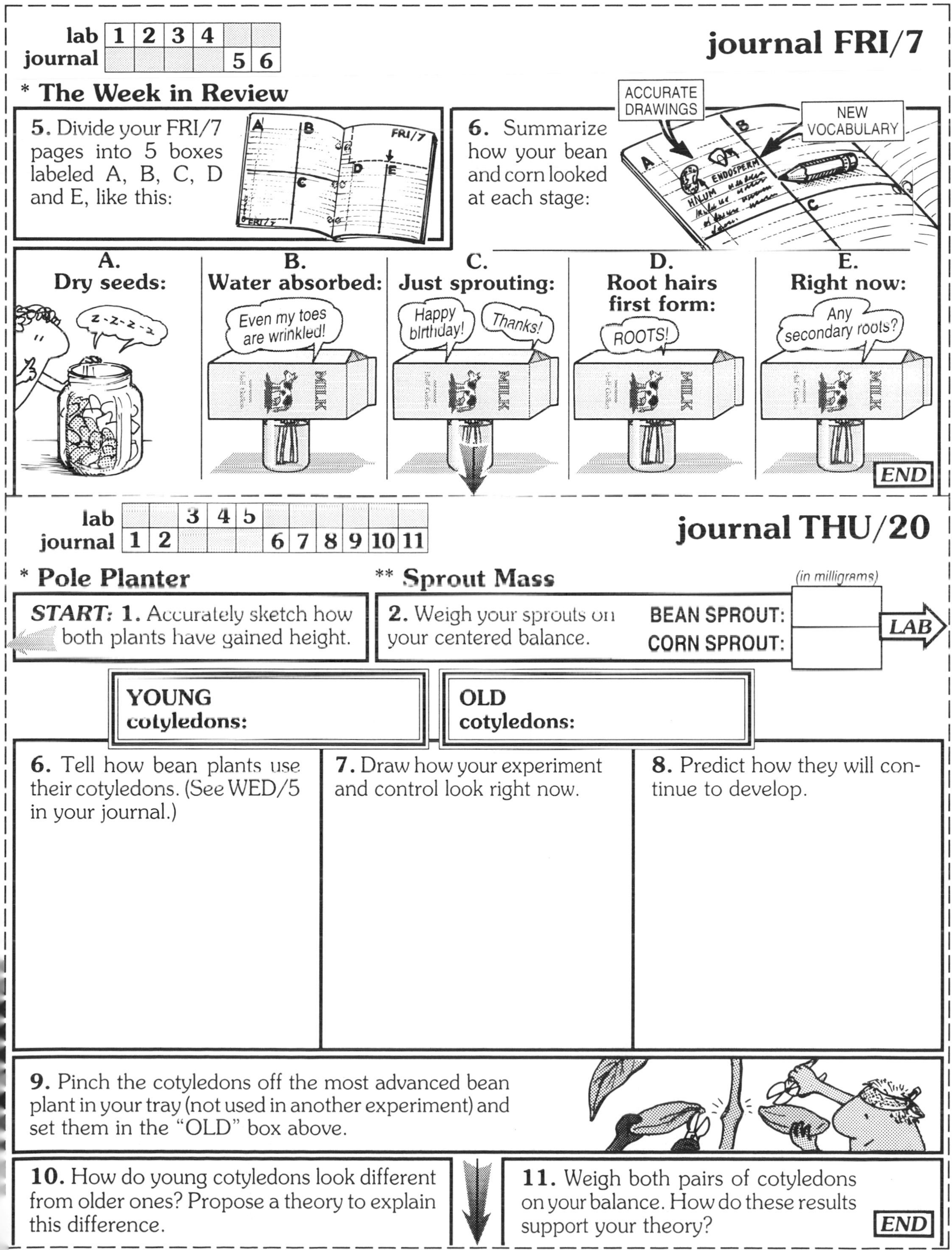

lab	1	2	3	4		
journal					5	6

journal FRI/7

* The Week in Review

5. Divide your FRI/7 pages into 5 boxes labeled A, B, C, D and E, like this:

6. Summarize how your bean and corn looked at each stage:

A. Dry seeds:

B. Water absorbed:

C. Just sprouting:

D. Root hairs first form:

E. Right now:

END

lab			3	4	5						
journal	1	2				6	7	8	9	10	11

journal THU/20

* Pole Planter

START: **1.** Accurately sketch how both plants have gained height.

** Sprout Mass

2. Weigh your sprouts on your centered balance.

	(in milligrams)
BEAN SPROUT:	
CORN SPROUT:	

LAB

YOUNG cotyledons:

OLD cotyledons:

6. Tell how bean plants use their cotyledons. (See WED/5 in your journal.)

7. Draw how your experiment and control look right now.

8. Predict how they will continue to develop.

9. Pinch the cotyledons off the most advanced bean plant in your tray (not used in another experiment) and set them in the "OLD" box above.

10. How do young cotyledons look different from older ones? Propose a theory to explain this difference.

11. Weigh both pairs of cotyledons on your balance. How do these results support your theory?

END

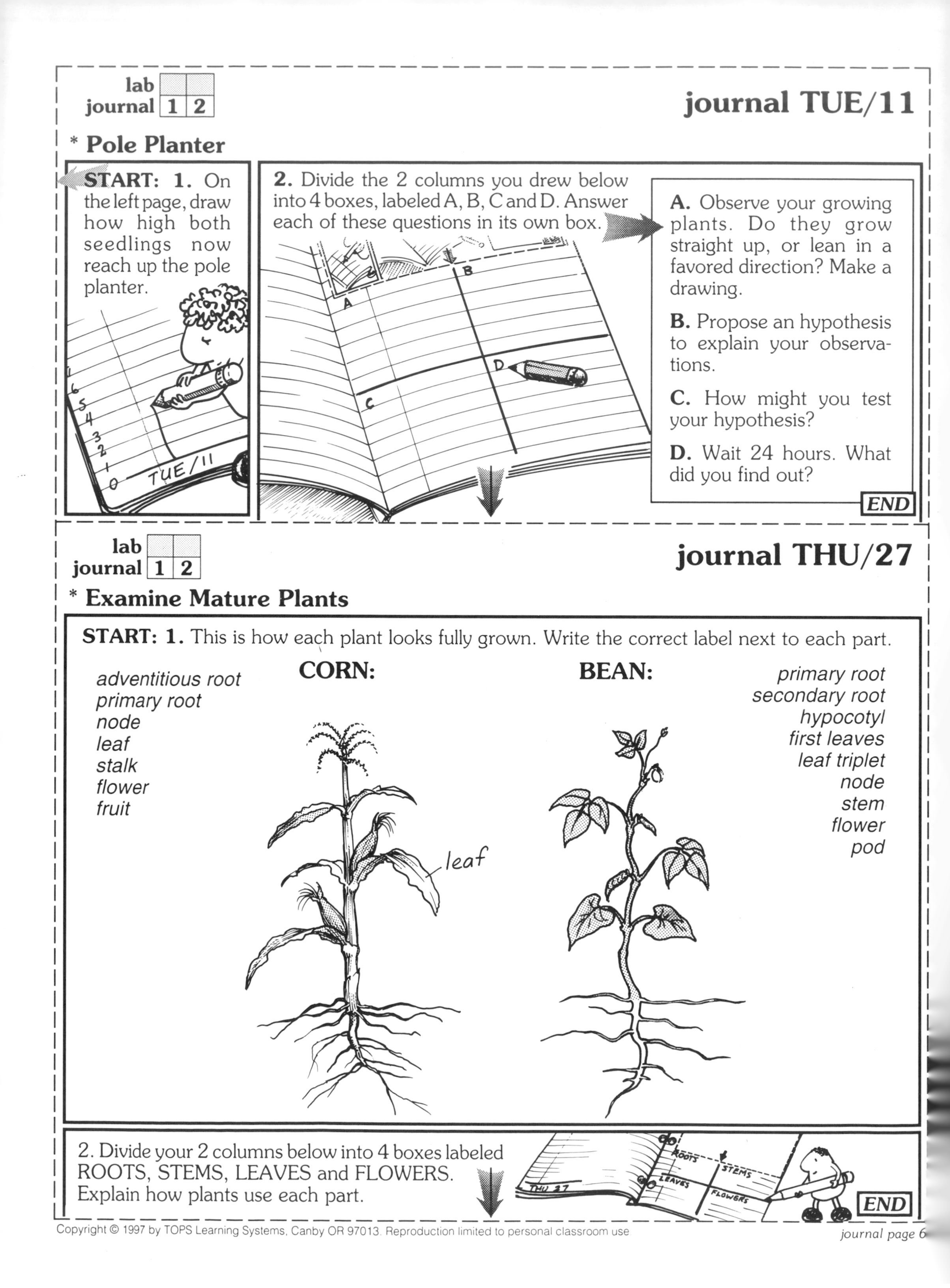

lab journal 1 2

journal TUE/11

* Pole Planter

START: 1. On the left page, draw how high both seedlings now reach up the pole planter.

2. Divide the 2 columns you drew below into 4 boxes, labeled A, B, C and D. Answer each of these questions in its own box.

A. Observe your growing plants. Do they grow straight up, or lean in a favored direction? Make a drawing.

B. Propose an hypothesis to explain your observations.

C. How might you test your hypothesis?

D. Wait 24 hours. What did you find out?

END

lab journal 1 2

journal THU/27

* Examine Mature Plants

START: 1. This is how each plant looks fully grown. Write the correct label next to each part.

CORN:

adventitious root
primary root
node
leaf
stalk
flower
fruit

BEAN:

primary root
secondary root
hypocotyl
first leaves
leaf triplet
node
stem
flower
pod

2. Divide your 2 columns below into 4 boxes labeled ROOTS, STEMS, LEAVES and FLOWERS. Explain how plants use each part.

END

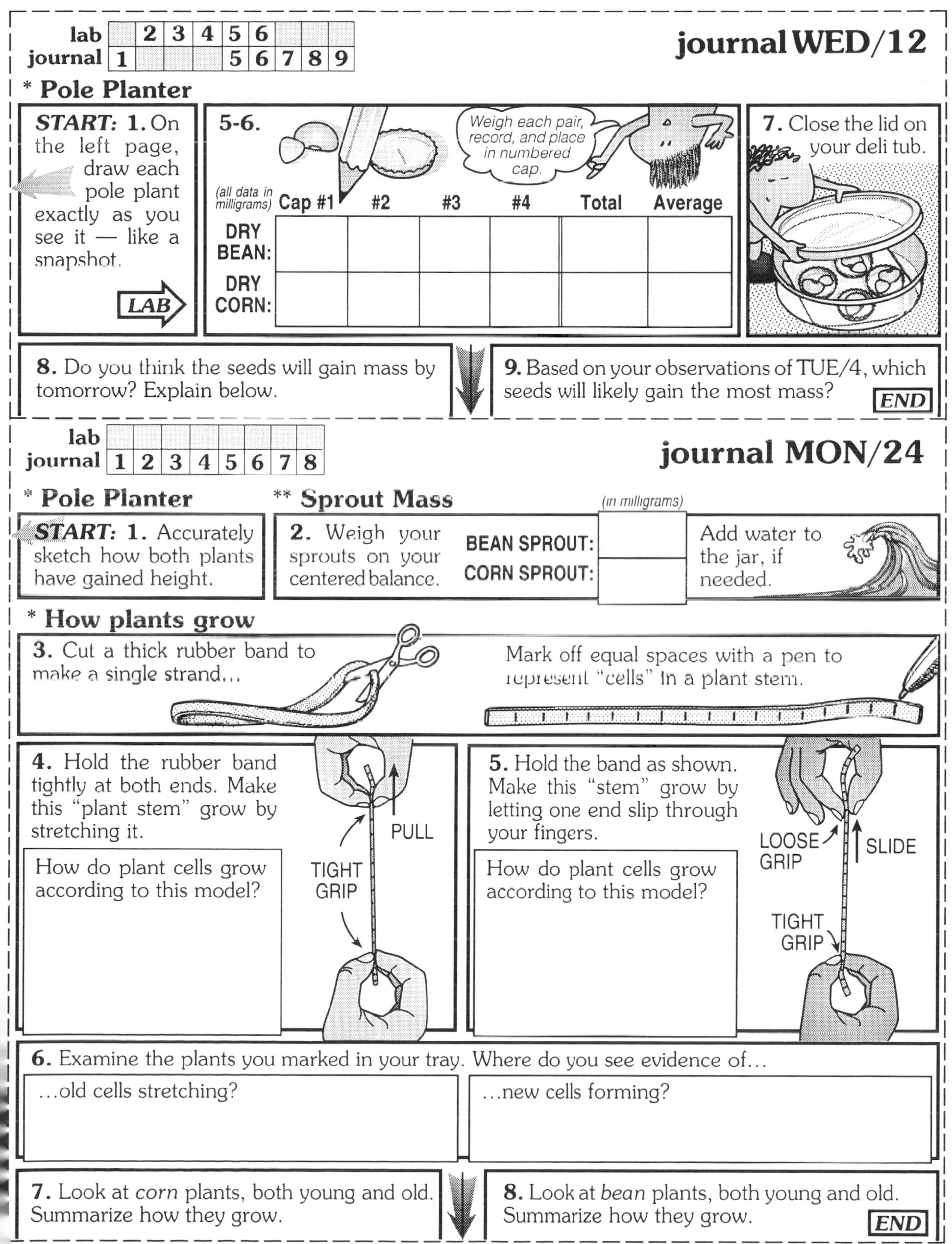

lab		2	3	4	5	6			
journal	1				5	6	7	8	9

journal WED/12

* Pole Planter

***START:* 1.** On the left page, draw each pole plant exactly as you see it — like a snapshot.

LAB

5-6.

(all data in milligrams)

	Cap #1	#2	#3	#4	Total	Average
DRY BEAN:						
DRY CORN:						

7. Close the lid on your deli tub.

8. Do you think the seeds will gain mass by tomorrow? Explain below.

9. Based on your observations of TUE/4, which seeds will likely gain the most mass?

END

lab								
journal	1	2	3	4	5	6	7	8

journal MON/24

* Pole Planter

***START:* 1.** Accurately sketch how both plants have gained height.

** Sprout Mass

2. Weigh your sprouts on your centered balance.

(in milligrams)	
BEAN SPROUT:	
CORN SPROUT:	

Add water to the jar, if needed.

* How plants grow

3. Cut a thick rubber band to make a single strand...

Mark off equal spaces with a pen to represent "cells" in a plant stem.

4. Hold the rubber band tightly at both ends. Make this "plant stem" grow by stretching it.

How do plant cells grow according to this model?

5. Hold the band as shown. Make this "stem" grow by letting one end slip through your fingers.

How do plant cells grow according to this model?

6. Examine the plants you marked in your tray. Where do you see evidence of...

...old cells stretching?

...new cells forming?

7. Look at *corn* plants, both young and old. Summarize how they grow.

8. Look at *bean* plants, both young and old. Summarize how they grow.

END

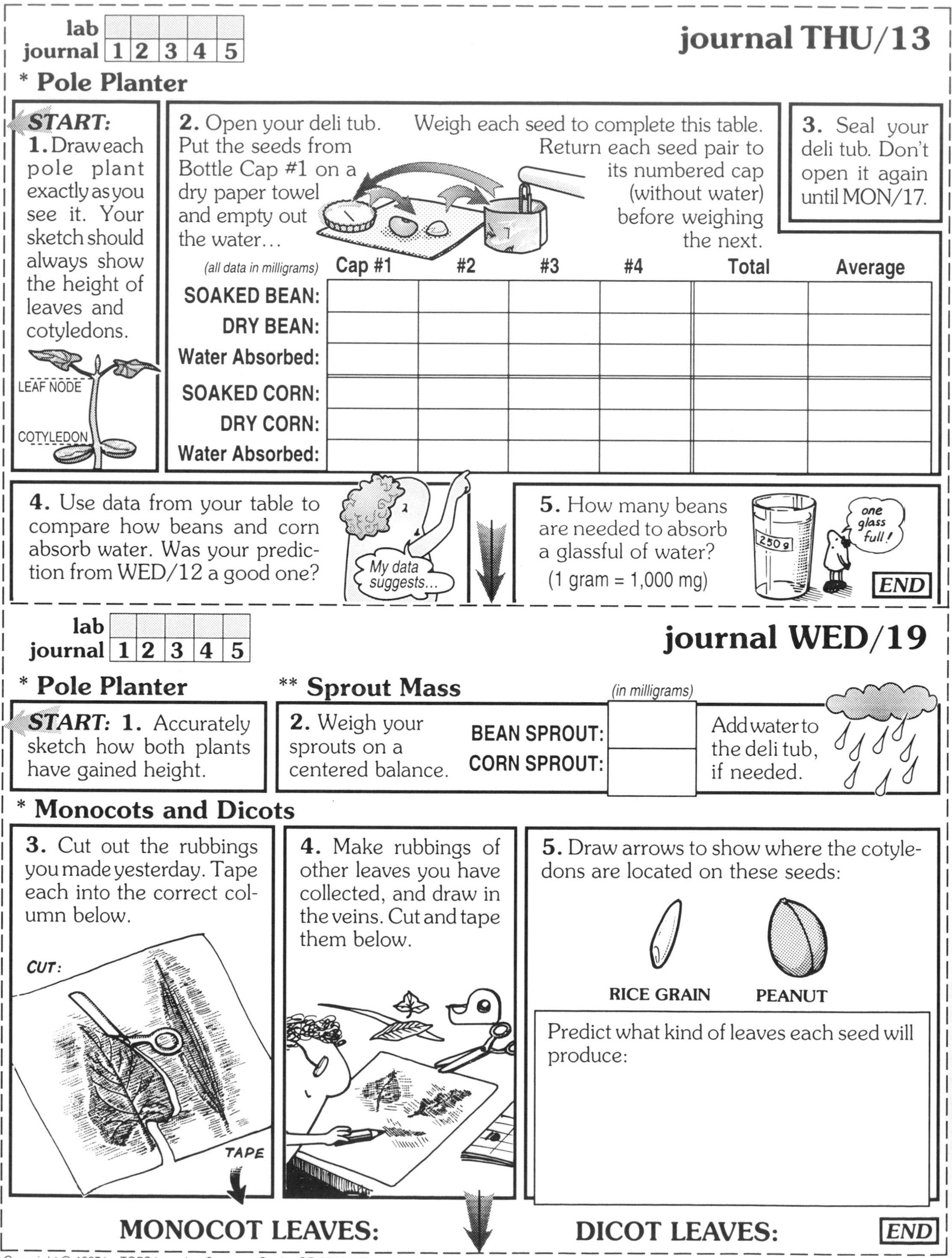

lab
journal 1 2 3 4 5

journal THU/13

* **Pole Planter**

START: **1.** Draw each pole plant exactly as you see it. Your sketch should always show the height of leaves and cotyledons.

2. Open your deli tub. Put the seeds from Bottle Cap #1 on a dry paper towel and empty out the water... Weigh each seed to complete this table. Return each seed pair to its numbered cap (without water) before weighing the next.

(all data in milligrams)	Cap #1	#2	#3	#4	Total	Average
SOAKED BEAN:						
DRY BEAN:						
Water Absorbed:						
SOAKED CORN:						
DRY CORN:						
Water Absorbed:						

3. Seal your deli tub. Don't open it again until MON/17.

4. Use data from your table to compare how beans and corn absorb water. Was your prediction from WED/12 a good one?

5. How many beans are needed to absorb a glassful of water? (1 gram = 1,000 mg)

END

lab
journal 1 2 3 4 5

journal WED/19

* **Pole Planter**

START: 1. Accurately sketch how both plants have gained height.

** **Sprout Mass**

2. Weigh your sprouts on a centered balance.

	(in milligrams)
BEAN SPROUT:	
CORN SPROUT:	

Add water to the deli tub, if needed.

* **Monocots and Dicots**

3. Cut out the rubbings you made yesterday. Tape each into the correct column below.

4. Make rubbings of other leaves you have collected, and draw in the veins. Cut and tape them below.

5. Draw arrows to show where the cotyledons are located on these seeds:

RICE GRAIN PEANUT

Predict what kind of leaves each seed will produce:

MONOCOT LEAVES: **DICOT LEAVES:**

END

lab		2	3	4	5	6	7		
journal	1							8	9

journal FRI/14

* **Pole Planter**

START: **1.** Accurately draw on the left page how both plants have gained height. **LAB**

8. Leaves turn green in sunlight as they use the sun's energy to photosynthesize their own food.

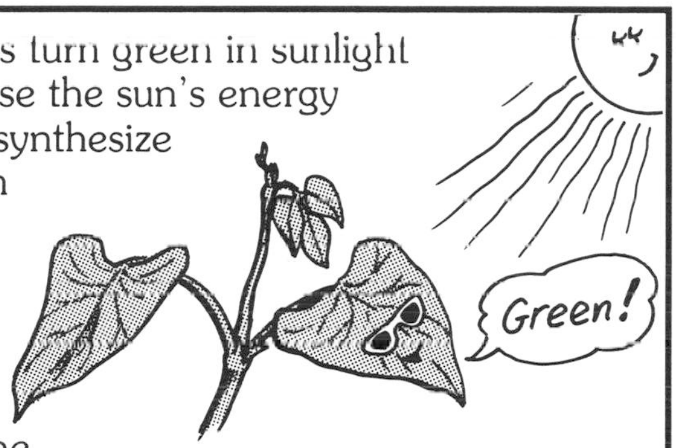

Predict the color of your leaf after being covered for a long time with foil. Explain your reasons.

9. Leaves breathe air through tiny openings (stomata) in their leaves. Stomata are too small to see without a microscope. If grease plugs these holes, the leaves can't breathe.

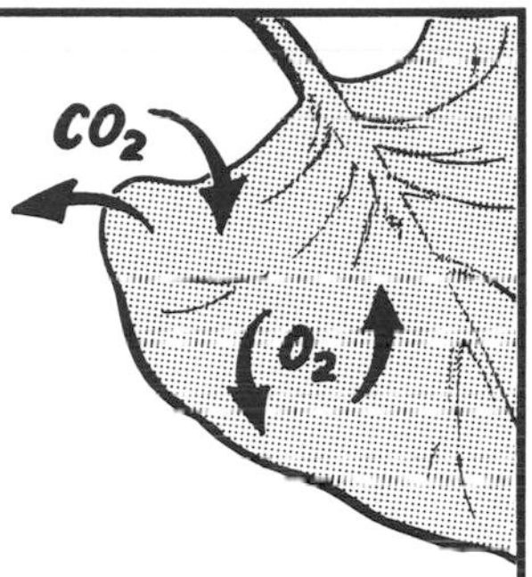

How might this experiment tell us where the stomata are located? **END**

lab		2	3	4			
journal	1			4	5	6	7

journal MON/17

* **Pole Planter**

START: **1.** Draw accurately how both plants have grown. **LAB**

4. Complete this mass table. Be sure to center your balance first…

(all data in milligrams)		DRY Wed/12	SOAKED Thu/13	SPROUTED Mon/17
BEAN:	#			
CORN:	#			

5. Clean out your deli tub, and line it with 3 clean, moist towels. Keep both weighed sprouts inside with the lid closed.

6. How much *more* mass did each seed gain after the 24-hour soak of THU/13?

7. What *other* way can seedlings gain mass besides absorbing water? **END**

PAPER MASS:

1000 mg (1 gram)

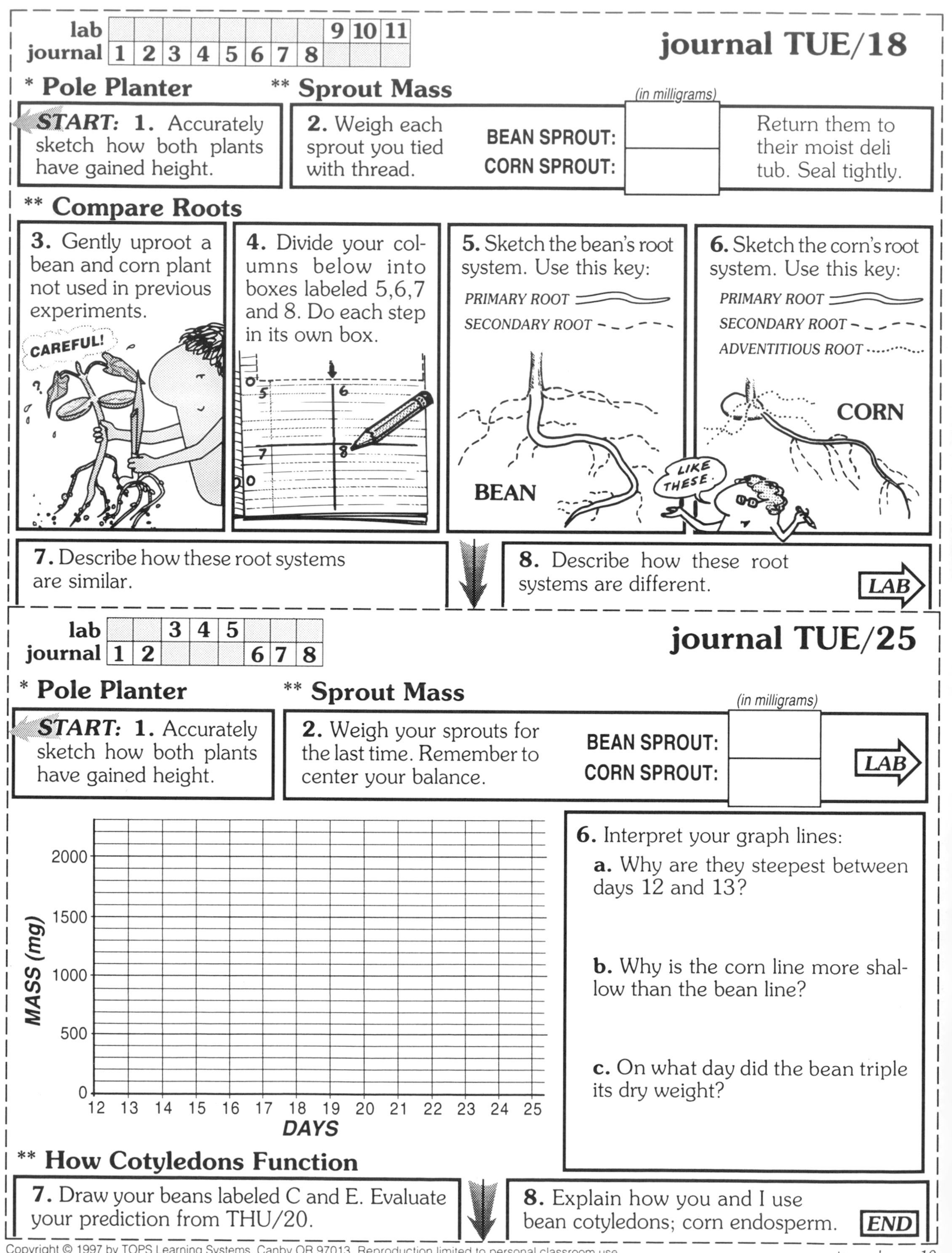

lab									9	10	11
journal	1	2	3	4	5	6	7	8			

journal TUE/18

* Pole Planter

START: 1. Accurately sketch how both plants have gained height.

** Sprout Mass

2. Weigh each sprout you tied with thread.

(in milligrams)

BEAN SPROUT:

CORN SPROUT:

Return them to their moist deli tub. Seal tightly.

** Compare Roots

3. Gently uproot a bean and corn plant not used in previous experiments.

4. Divide your columns below into boxes labeled 5,6,7 and 8. Do each step in its own box.

5. Sketch the bean's root system. Use this key:

6. Sketch the corn's root system. Use this key:

7. Describe how these root systems are similar.

8. Describe how these root systems are different.

LAB

lab			3	4	5			
journal	1	2				6	7	8

journal TUE/25

* Pole Planter

START: 1. Accurately sketch how both plants have gained height.

** Sprout Mass

2. Weigh your sprouts for the last time. Remember to center your balance.

(in milligrams)

BEAN SPROUT:

CORN SPROUT:

LAB

6. Interpret your graph lines:

a. Why are they steepest between days 12 and 13?

b. Why is the corn line more shallow than the bean line?

c. On what day did the bean triple its dry weight?

** How Cotyledons Function

7. Draw your beans labeled C and E. Evaluate your prediction from THU/20.

8. Explain how you and I use bean cotyledons; corn endosperm.

END

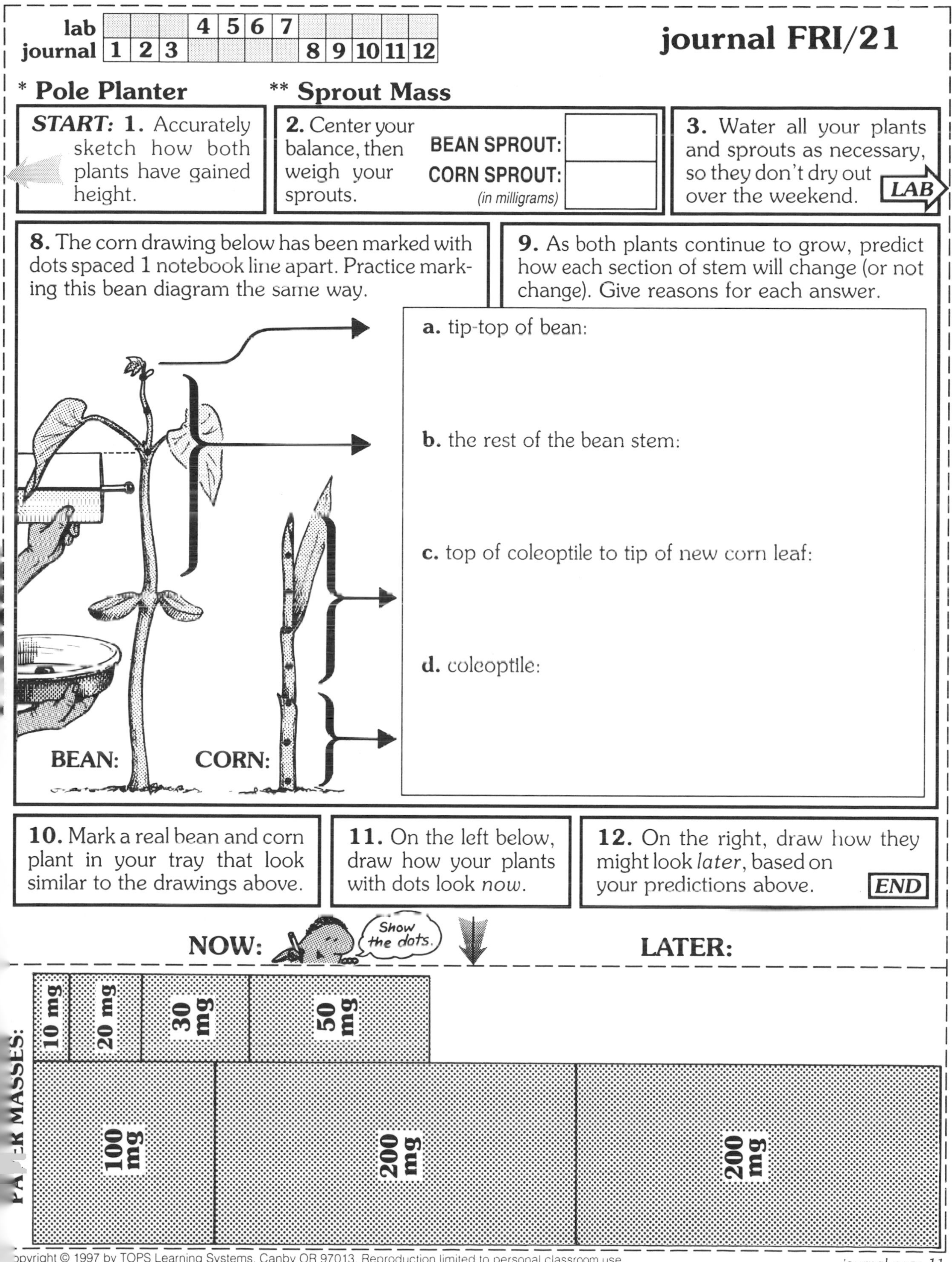

lab | 4 | 5 | 6 | 7
journal | 1 | 2 | 3 | 8 | 9 | 10 | 11 | 12

journal FRI/21

* Pole Planter ** Sprout Mass

START: 1. Accurately sketch how both plants have gained height.

2. Center your balance, then weigh your sprouts.

BEAN SPROUT:	
CORN SPROUT: *(in milligrams)*	

3. Water all your plants and sprouts as necessary, so they don't dry out over the weekend. LAB

8. The corn drawing below has been marked with dots spaced 1 notebook line apart. Practice marking this bean diagram the same way.

9. As both plants continue to grow, predict how each section of stem will change (or not change). Give reasons for each answer.

a. tip-top of bean:

b. the rest of the bean stem:

c. top of coleoptile to tip of new corn leaf:

d. coleoptile:

10. Mark a real bean and corn plant in your tray that look similar to the drawings above.

11. On the left below, draw how your plants with dots look *now*.

12. On the right, draw how they might look *later*, based on your predictions above. END

NOW:

LATER:

PAPER MASSES:

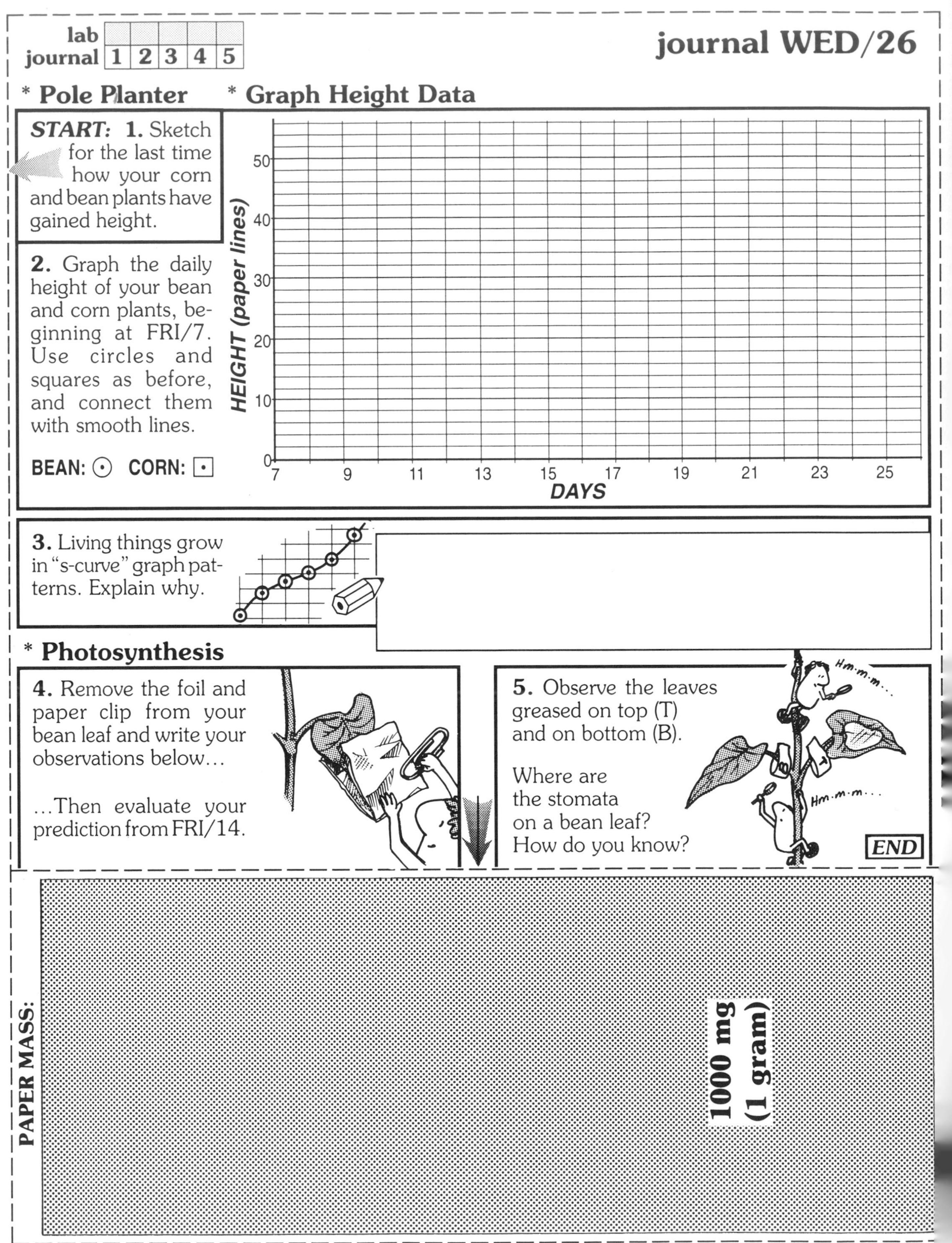

lab journal 1 2 3 4 5

journal WED/26

* Pole Planter

START: 1. Sketch for the last time how your corn and bean plants have gained height.

2. Graph the daily height of your bean and corn plants, beginning at FRI/7. Use circles and squares as before, and connect them with smooth lines.

BEAN: ⊙ **CORN:** ⊡

* Graph Height Data

3. Living things grow in "s-curve" graph patterns. Explain why.

* Photosynthesis

4. Remove the foil and paper clip from your bean leaf and write your observations below…

…Then evaluate your prediction from FRI/14.

5. Observe the leaves greased on top (T) and on bottom (B).

Where are the stomata on a bean leaf? How do you know?

END

PAPER MASS:

1000 mg (1 gram)